INSTRUCTOR'S MANUAL FOR

# Ecosystem Management

## About Island Press

Island Press is the only nonprofit organization in the United States whose principal purpose is the publication of books on environmental issues and natural resource management. We provide solutions-oriented information to professionals, public officials, business and community leaders, and concerned citizens who are shaping responses to environmental problems.

In 2002, Island Press celebrates its eighteenth anniversary as the leading provider of timely and practical books that take a multidisciplinary approach to critical environmental concerns. Our growing list of titles reflects our commitment to bringing the best of an expanding body of literature to the environmental community throughout North America and the world.

Support for Island Press is provided by The Nathan Cummings Foundation, Geraldine R. Dodge Foundation, Doris Duke Charitable Foundation, Educational Foundation of America, The Charles Engelhard Foundation, The Ford Foundation, The George Gund Foundation, The Vira I. Heinz Endowment, The William and Flora Hewlett Foundation, Henry Luce Foundation, The John D. and Catherine T. MacArthur Foundation, The Andrew W. Mellon Foundation, The Moriah Fund, The Curtis and Edith Munson Foundation, National Fish and Wildlife Foundation, The New-Land Foundation, Oak Foundation, The Overbrook Foundation, The David and Lucile Packard Foundation, The Pew Charitable Trusts, The Rockefeller Foundation, The Winslow Foundation, and other generous donors.

INSTRUCTOR'S MANUAL FOR

# Ecosystem Management

## *Adaptive, Community-Based Conservation*

LARRY A. NIELSEN
GARY K. MEFFE
RICHARD L. KNIGHT

ISLAND PRESS
Washington • Covelo • London

ISBN 1-55963-825-7

Printed on recycled, acid-free paper

Manufactured in the United States of America

09 08 07 06 05 04 03 02 8 7 6 5 4 3 2 1

# Contents

# An Overview of the Book

## *Purpose of the Course and Overall Philosophy*

WE HOPE YOU WILL FIND THIS BOOK A USEFUL TOOL FOR ENGAGING STUDENTS IN THE roles and techniques of adaptive ecosystem management and community-based conservation. We created the book to match our perspectives about ecosystem management and about teaching.

We think of ecosystem management as the natural evolution of our ability to live in a place without spoiling it, to paraphrase Aldo Leopold. We also believe that our ability to live well on the land entails not only a strong sense of ecological principles, but also a strong commitment to the people and communities in which we live. Segregating nature from humans is a strategy that typically backfires in the long run. Therefore, we believe that a course on ecosystem management must blend aspects of biology, ecology, and sociology. Although we expect most courses that would use our book to be taught in departments of fisheries and wildlife, natural resources, or the like, we believe that the course must be integrative, with substantial coverage of the human dimensions.

We also believe that teaching should be active and collaborative, involving students as fully and as regularly as possible. This is true for any sort of learning, but it is especially important for a course that hopes to convey an appreciation for collaboration as a way to make decisions on the land. If our students are to become the stewards of community-based conservation, then we must teach them to do so by letting them participate in and take responsibility for their education.

We hope that the students participating in a course for which our book is used will come from a wide variety of backgrounds and interests. This diversity not only will make the learning environment richer, but it will also more fully reflect the actual situations in which ecosystem management will be applied—a richly diverse landscape of people with different but overlapping values, interests, and expectations.

## *Purpose of the Book*

Because of these beliefs, we have crafted a book that may look a bit different than others and that we expect to be used differently. Rather than a storehouse of knowledge needed to solve ecosystem management dilemmas, the book is an aid to learning. We hope it will provide enough information and stimulation to get students and instructors working together to develop ways of thinking that go far beyond the book's pages.

We have fought the temptation to make the book an encyclopedia of everything we know and need to know to approach conservation as ecosystem management. Such a book would be too heavy on detail and correspondingly light on integration. The other option was to make a book that dealt only with integration and application, containing no fundamental information about biology or sociology. We rejected this approach as well, thinking that we wanted to provide a strong foundation of basic principles in ecosystem thinking so the book and course would be accessible to people with only some background in the technical aspects of biology, ecology, or sociology.

Consequently, the book contains some biological and sociological information, but it is tailored to represent those areas closest to the topic of ecosystem management. However, the book is small, given the topic it purports to cover, and we fully expect it to be only the starting point for learning about ecosystem management. The onus, therefore, is on the instructor and students to fill in the gaps—and the gaps are large.

## Overall Strategy of the Book

We have built the book around three ideas. First, we present content that may be familiar, but we hope it provides an extension of knowledge. For example, we have moved beyond population analysis to metapopulation analysis and beyond strategic planning to strategic management as a whole.

Second, we ask that readers use the information in new contexts. These contexts include cross-boundary management, community-based management, and adaptive management.

Third, we have provided landscape scenarios—imaginary ecosystems—where the new content and new context can be applied in a safe setting. By providing scenarios that have a rich texture of ecological and sociological information, we hope that readers and learners will be able to try out their ideas without the constraints that real places might have. A detailed explanation of the scenarios and how they might be used is included in the following chapter.

A one-semester or one-quarter course based on this book probably is best taught through two classroom meetings per week (about 1 hour each) and a longer "laboratory" period in which the larger problems are worked out, detailed discussions are held, or supplementary material is explored. We suggest that chapter readings be assigned before each classroom meeting, so the substantive materials are assimilated before class. The class time can then be spent on clarification of materials, supplemental lecture and discussion, and work on the shorter exercises. We anticipate that such a course would be most appropriate as a senior-level capstone course or as a graduate-level course.

## Using the Scenarios

The scenarios are integral to the book, and we strongly recommend that they be used as the basis for application in the course. We have constructed the book so that examples in the text and many of the exercises are keyed directly to the scenarios.

We recommend that the course begin with a journey through one or more of the scenarios, so that students become familiar with them right at the beginning. Thus, we introduce the landscape scenarios in Chapter 1. However, the scenarios are imaginary, so both instructors and students should feel free to supplement the material in the scenarios with other materials that enhance their learning.

## *Contributed Essays*

The book includes several essays written by practitioners of ecosystem management. Designed to provide additional perspective on the topic, the essays also demonstrate to students how some people are actually using ecosystem management principles—providing testimonials to the reality that students will encounter.

The essays describe how land managers are using the concepts of ecosystem management to accomplish good results on the land. Therefore, the essays can be read and used at any time in the course, not just when the materials in the corresponding chapter have been covered. For example, if a guest speaker is visiting the campus and will be talking about a topic covered in one of the essays, the essay might be read as preparation for the visit.

The essays can also be the basis for supplementary activities by students. For example, students could build on The Nature Conservancy essay by contacting the local TNC office and finding out how business is done there. Several questions at the end of each essay are intended as the springboards toward further discussion and analyses of the topic. We encourage classes to address these in open discussion.

## *Text Boxes*

The book includes many boxes to supplement the main text. They provide examples, perspectives, definitions, summaries, or other aids to learning. We recommend that the boxes be integrated into the class as fully as possible, and subsequent chapters in this Instructor's Manual include some ideas for building on materials in the boxes.

## *Web Sites*

We hope this book will stimulate students to find out much more about the topics and places we have covered. This is particularly important because the material changes so rapidly. Therefore, we encourage students and instructors to use the Internet to supplement materials in the book. We have not embedded Web site addresses in the text, because they can change so unpredictably. In subsequent chapters of this manual, we have included many Web sites that link to organizations and topics covered in the text. The listing of a Web site in this manual, however, is neither an endorsement of the site by the authors nor a representation by the authors that the materials at the site are accurate, up-to-date, or complete. As always, materials available on the Internet must be evaluated personally before use.

## *The Different Types of Exercises*

The book includes three types of exercises, designed for different purposes and for different settings.

1. *Think About It!* Think-about-it exercises are designed to open the student's mind in preparation for receiving the material in the chapter. Often the exercises introduce a topic by asking the student to reflect on how the topic has been encountered in his or her life. The exercises can be done individually or in small groups. They usually should take no more than a few minutes; some interaction among students may help get them involved in the exercise and set the stage for subsequent interactions.
2. *Talk About It!* Talk-about-it exercises represent an array of activities, encompassing the more traditional forms of problem solving and group interaction. This style of exercise is designed to engage students in the details of the material; consequently, these exercises should be used in direct conjunction with the material as it is presented. They are designed to be conducted in small groups, and most should take from a few minutes to a portion of a discussion or laboratory period. Some talk-about-it exercises use the scenarios as a base and therefore will require extensive exposure to the scenarios before they are performed.
3. *Collaborate on It!* Collaborate-on-it exercises are the most important exercises in the book, representing the major device for getting students actively involved in ecosystem management. These exercises primarily use the scenarios as a base, so extensive exposure to the scenarios will be necessary before the exercises are performed. The exercises are usually quite involved, asking students to perform at higher levels of thinking—application, analysis, synthesis, and evaluation. The exercises also frequently ask students to go beyond the information presented in the book to develop new perspectives on a particular situation or on ecosystem management in general.

   Collaborate-on-it exercises are designed to be conducted in small groups, with the expectation that ample opportunity will be provided for interaction among groups. They may require much or all of a discussion or laboratory period, and several may require extensive out-of-class work, extending over several class periods. These exercises become more comprehensive as the book proceeds, with the major collaboration exercises occurring in Chapter 11. Thus, the use of these exercises must be built carefully into the structure of the course.

As you will come to see, this is a rigorous (but fun!) course based on an ambitious approach to problem solving in natural resource management. It is also a different way of teaching and learning. We are convinced that the "lecture/take notes/repeat-it-later" approach to learning is becoming outdated for future managers of our natural resources. The reality of real-world situations is too complex, fuzzy, and contentious to rely simply on a theoretical and empirical knowledge of biology, ecology, and various other sciences. The real challenges for practitioners are in the human dimensions, and the classroom is the time and place to begin to train our future professionals to deal with the messy realities of life. We think this book, and a course based upon it, is a good start. We solicit and welcome feedback from instructors based on your experiences in the classroom, so we can learn from you and make improvements in future editions of the book.

# The Landscape Scenarios

CHAPTER 1 CONTAINS THREE LANDSCAPE SCENARIOS THAT ARE AT THE HEART OF THE book. The scenarios provide a different context in which learning can occur, and they represent a major departure from other textbooks.

## *Purpose of the Scenarios*

The three scenarios create imaginary and complex but realistic ecosystems in which students can apply the lessons of ecosystem management. The purpose is to provide a focused example that is fairly complete and fully available for students to use throughout the course. Although virtually any ecosystem could be used as the basis for ecosystem management (that is the purpose of the concept), gathering the necessary information would be a very substantial task. Therefore, we have compiled the information for you.

We have chosen to use imaginary ecosystems rather than real ones for several reasons. First, the information in real ecosystems changes so fast that it would be out of date for class use almost immediately. Second, teachers or students may have experience with real ecosystems that would not allow them to treat the exercises with complete objectivity. Third, using imaginary ecosystems lets us insert as many interesting situations as we like; we are not limited by the need for strict accuracy.

The scenarios let students focus their attention on one or a few situations throughout the course, rather than having to work with different situations for all the concepts. Consequently, they should be able to develop an in-depth knowledge and feeling for the material and its application in the context of a particular scenario.

The scenarios are designed to allow the course to be student-centered rather than teacher-centered. By providing a fully developed example in which students can work, the focus of a course can be on application and discovery, rather than on lectures and information transmission. As mentioned earlier, scenario-based exercises (usually the collaborate-on-it exercises) are distributed throughout the textbook and are intended to be the vehicle through which most learning occurs.

## *Content of the Scenarios*

The places, people, and situations in the landscape scenarios are fictional, created by the authors. As works of fiction state at the beginning, any resemblance to real persons, places, or events are strictly coincidental. However, the scenarios are intended to be realistic. Aspects in the scenarios are based on real situations with which the authors are familiar.

The biological information in the scenarios is a mixture of real and imaginary. For example, some species are real species (e.g., sandhill cranes, gopher tortoise), whereas others are fictional. Some of the biological characteristics of real species have been modified to fit the situation, and some biological information has been changed (e.g., a species may have been moved from threatened to endangered or to de-listed). Also, the species descriptions, even for real species, are very brief. We encourage students to investigate real species to update the information, rather than rely only on the data in the book.

The Round Lake ecosystem is meant to resemble an ecosystem in the northeastern or upper midwestern United States. The Snow River ecosystem is meant to resemble an intermountain West ecosystem. The Paumassee-Dee-Queen (PDQ) ecosystem is meant to resemble a southeastern United States ecosystem. Round Lake and Snow River are similar in their content and scope. The ROLE Model (Round Lake) is more oriented toward aquatic issues, whereas SnowPACT (Snow River) is more oriented to terrestrial issues. The PDQ ecosystem, however, is much larger in area, comprising a multiple river region, and is more loosely organized.

## *How to Use the Scenarios*

The scenarios should be the basis and focus of teaching and learning throughout the course. We have developed the scenarios to be useful as tools for active learning, primarily through the exercises, but also as information to use as examples for all the material in the book.

We have included three different scenarios to allow instructors two basic ways to organize classes, as follows:

1. One scenario can be used throughout the class, with all students working on just that scenario throughout the term. For group work, the class is divided into teams that compare their results based on the same scenario.

The advantage to this approach is that everyone becomes familiar with the same ecosystem, and confusion from reporting on different ecosystems is avoided.

2. The class can be divided into two or three groups, each of which works with one of the scenarios throughout the term. This approach allows each group to become familiar with one scenario, but each group can report orally so that all groups see the differences and similarities among the three scenarios. The advantage to this approach is the diversity of experience that results from using more than one system, though it is trickier to keep all the information straight.

We have placed the scenarios as the first chapter because we believe students should read about and "inhabit" their scenario first, before starting work on any of the substantive material. Therefore, we suggest that Chapter 1 be assigned immediately at the beginning of the term, followed by a discussion of the content of the scenario. We recommend a laboratory session to answer these and other questions:

- How would you characterize this ecosystem?
- What are the major species, populations, and biological communities of interest?
- Who are the interesting people and groups in the ecosystem?
- What are special resources available in this ecosystem? Major problems that exist now? Likely points of issue?
- How did the ecosystem get to the state it is in now?
- Would you like to be living and working in this ecosystem? Why or why not?

We also recommend that students begin to develop a small journal with their thoughts about how the material relates to the scenario as they go along. The technical chapters (Chapters 5–9) are less closely tied to the scenarios than are the stakeholder and process chapters (Chapters 10–12), so students will need to be aware of the material in the technical chapters as they go along. In later chapters, the exercises assume that the students will be applying technical aspects to the process work. This journal can also be used as a supplementary evaluation tool in the class.

## *Supplementing the Scenarios*

The scenarios can be supplemented in various ways. First, other information can be added. This is especially useful for biological aspects. Asking students to develop more information for species, populations, and communities based on the technical chapters would help build the scenario base for subsequent process work. Assignments or special studies for interested or more advanced students taking the course may be given.

Second, other examples or cases can be added to the exercises in the textbook that relate to scenarios. In many scenario exercises, a specific topic has been indicated in the exercise. However, replacing that topic or adding others may make more sense based on how the course is taught and what is emphasized.

Third, entirely new scenarios can be added. Students who desire special studies can be asked to develop a new scenario along the lines of those included, but specific to the region of the country in which the course is being taught.

Fourth, the scenarios can be used for role-playing exercises, which is a very useful learning device. However, because instructors vary widely in their opinions about role playing, we have avoided building the exercises around role playing.

A final word about the scenarios. It is our experience that they provide a very rich experience for learners and are the basis for everything else that follows. Dealing with the complexities, nuances, and realities of these places will teach students more than any theoretical treatment of conservation can possibly do. Thus, we encourage students and instructors alike to mentally become part of these places, to feel like inhabitants, and to care about the habitats, species, and people encountered there. Doing so will result in a much more rewarding learning experience.

CHAPTER 2

# Getting a Grip on Ecosystem Management

## *Learning Objectives for the Chapter*

THIS CHAPTER PROVIDES STUDENTS WITH THE FOUNDATION FOR UNDERSTANDING AND working with ecosystem management through the remainder of the book. Specific learning objectives include:

- Understanding that ecosystem management is the result of an evolutionary process, naturally leading from collective past experiences—both successes and failures—in natural resource management, rather than a sudden revolutionary change.
- Understanding how ecosystem management contrasts with more traditional modes of management.
- Understanding why the application of command and control to highly complex and unpredictable natural systems typically fails in the long run, in what has been called "the pathology of natural resource management." This leads to the ecosystem approach.
- Introduction to a conceptual model of ecosystem management that will be the focus for the entire book.
- Defining and understanding what is meant by ecosystem management, including core concepts such as looking beyond political boundaries, involving stakeholders in decision making, having a long-term perspective, and defining future conditions.
- Beginning the process of working with exercises that challenge the student to deal with complex landscapes and issues.

## *Major Points of the Chapter*

The chapter tries to get students to think outside of the traditional management box and to understand that past efforts to tightly control managed systems in a prescriptive and piecemeal manner, by narrowly focused management agencies, do not work over the long haul. We also want students to realize that effective management involves *much* more than the biological and ecological knowledge that most of them focus on in their studies. It also requires developing skills and perspectives they may never have considered, such as "people" skills and viewpoints from the humanities. Ecosystem management is as much about dealing effectively with people as it is about scientific understanding, perhaps more so. This must be acknowledged and embraced by students if they are to become effective professionals in natural resource management.

## *Items That May Come Up While Teaching This Chapter*

Students may be surprised at the amount of "nonscientific" emphasis in our presentation of ecosystem management; this is good, and it should be used as a springboard for open discussion. The point needs to be made that management is not conducted in an academic or theoretical vacuum where ecological principles may be freely applied across the landscape. That landscape is in fact heavily peopled, and these people do not necessarily share the same scientific understanding, worldview, or concern for the environment as the students. Do we dismiss these folks and try to "do our thing" in spite of them? Or do we learn to understand them, their issues and concerns, their problems and perspectives, and work with them to find common and acceptable solutions?

Students may think that the environment has given up enough and that no more compromise is in order. This is an opportunity to discuss pragmatism versus idealism and what approaches will likely work best to protect remaining resources. Do we continue to produce clear winners and losers, or do we try new ways that may not get us everything they want, but will result overall in a habitable planet with functioning ecosystems?

Interesting discussions may ensue regarding command and control. Command and control have been very successful for many aspects of human existence to date, and in fact our civilization was built on controlling our environment. Perceptive students may bring this up. We are simply pointing out that the complexities and unpredictabilities of nature dictate (we now know) that tight control with consistent and effective prediction is impossible. Thus, understanding this and avoiding the need for tight and effective control will preclude placing us into management corners from which there is no escape. It is more effective in the long run (though often tougher in the short run) to understand natural variation and try to work with it rather than control it. Maintaining system resilience, rather than practicing tight control, is the goal.

Finally, older students, or those with more of a traditional wildlife and fisheries background, may resent and misunderstand comparisons of traditional and ecosystem approaches. Sensitivity is important here. We are building upon the knowledge and strengths of a century of management, not discarding it as wrong. Times change, understanding increases, and approaches evolve.

## *Supplementary Activities*

Discussion of the human elements of ecosystem management would help bring this topic to the forefront in the thinking of students. They may need to consider courses in the humanities or other fields that will broaden their training and abilities.

Use of state resource agency Web sites, or those of the U.S. Fish and Wildlife Service, the U.S. Forest Service, or other federal agencies, could bring in other perspectives of what ecosystem management is all about. Comparisons of such approaches and ideas would be constructive.

## *Conducting the Exercises*

### EXERCISE 2.1

#### How to Introduce or Conduct the Exercise

This is an informal, open class discussion. It might be most efficient for the instructor to have an issue in mind so you don't spend time debating what to discuss.

#### What to Expect When Students Do the Exercise

Depending on the example chosen and how familiar they are with it, students may have a difficult time finding aspects that fit either the traditional or the ecosystem model. You may need to prod them or have aspects ready to give them.

#### Desired Outcomes for the Exercise

An appreciation for how management might be changing to embrace broader perspectives and how society may be expecting more from management than the impasses and confrontations that defined so much of our recent history.

#### Points to Be Made Based on the Exercise

- We are in a transition period in management and have an opportunity to help define the future.
- Ecosystem management may be slower and more difficult up front to accomplish.
- Ecosystem management may be more acceptable to many segments of society.

Do not be concerned if these points do not arise; this is very early in the course, and it may be too optimistic to expect this level of understanding now.

## EXERCISE 2.2

### How to Introduce or Conduct the Exercise

It may be best to leave this to the individual students to ponder, and then seek a few volunteers to tell the class what they came up with.

### What to Expect When Students Do the Exercise

This is not a profound exercise, but merely an opportunity for students to relate back to their own world the ideas of resource waste and discovering better ways of doing things.

### Desired Outcomes for the Exercise

Realization by students that the past is not necessarily a map to the future; there are alternatives.

### Points to Be Made Based on the Exercise

- There usually are better ways to do things, and we can innovate.

## EXERCISE 2.3

### How to Introduce or Conduct the Exercise

Again, students should probably ponder this on their own, and then a few can share what they discovered.

### What to Expect When Students Do the Exercise

Students may have difficulty coming up with good examples; they are not used to thinking about such issues. Instructors may need to be ready to help them along.

### Desired Outcomes for the Exercise

One or two good examples that supplement the text and that perhaps have relevance close to home will suffice.

### Points to Be Made Based on the Exercise

- Command and control has a real cost, and there are alternative approaches.

## EXERCISE 2.4

### How to Introduce or Conduct the Exercise

This is best conducted as a class discussion, prompting students to think about and discuss what happens when riverine flows are controlled.

### What to Expect When Students Do the Exercise

Students should be able to come up with the basics of altered/controlled flows through damming, levy building, and dredging, and how that works as planned—most of the time.

### Desired Outcomes for the Exercise

Students need to understand that the unusual events—the 100- or 500-year floods—violate the assumptions of command and control, and bring us face-to-face with the pathology. Initial control of the system works; then focus is on flood control rather than normal riverine behavior, followed by heavy capitalization—industry and communities built in floodplains—which completes the pathology. When major flooding occurs and control fails, huge damages result. Eventual failure is built into the system at the outset.

### Points to Be Made Based on the Exercise

- This pathology is real and ultimately leads to management failure.
- Alternatives exist with some foresight. For example, floodplains could be used for wildlife corridors and some agricultural uses rather than cities. The reason floodplains are so fertile is that they flood! This fertility could be used by agriculture; if crops are lost once every few years, we are probably still ahead because of the renewed fertility and good crops the rest of the time.

## EXERCISE 2.5

### How to Introduce or Conduct the Exercise

This is another class discussion item that should be guided by the instructor.

### What to Expect When Students Do the Exercise

Students should be able to envision changes in management approach and resultant outcomes based on thinking about natural variation, maintaining resilience, and the implications of those changes for management.

### Desired Outcomes for the Exercise

Innovation on the part of students; thinking "outside the box" in coming up with new management approaches based on resilience.

### Points to Be Made Based on the Exercise

- Some foresight in thinking about how management in the context of natural variation can improve results.

## EXERCISE 2.6

### How to Introduce or Conduct the Exercise

This is the first exercise to use the scenarios. See the general description on using the scenario exercises. Be sure students have read their scenario before this point. It would be best to conduct this first one as a class discussion.

### What to Expect When Students Do the Exercise

This may start off slowly, as students are making their first foray into the scenario. Be patient, but be ready to draw them out with leading questions.

### Desired Outcomes for the Exercise

We would like to see students start to look beyond the map boundaries of reserves and begin to embrace ecological units. This is an important realization that will need to be carried on throughout the book. Similarly, they should begin to include a broad diversity of people into decision making and start to look at planning horizons of decades. They should begin to form at least rough ideas of how to do these things.

### Points to Be Made Based on the Exercise

- We need to start expanding the horizons in management along these three axes to be more effective.

## EXERCISE 2.7

### How to Introduce or Conduct the Exercise

See the general description on using the scenario exercises. This experience works well by counting out by fives or otherwise quickly assembling five groups. Assign one phrase per group, and give the groups 15 minutes or so to work, per the directions.

### What to Expect When Students Do the Exercise

Because students will be new to these concepts, they may be slow to develop some of these points. Instructor oversight and prodding may help.

### Desired Outcomes for the Exercise

This is an important exercise because it clearly indicates pros and cons for ecosystem management, and helps to further delineate it from more traditional approaches. And the best part is, the students themselves come up with the pros and cons.

### Points to Be Made Based on the Exercise

- Ecosystem management is initially slow and takes a great deal of front-end planning and relationship building.
- Traditional approaches work best in emergencies or when there are very specific needs that clearly can be met with a standard project.
- Ecosystem management is limited by time, money, institutional momentum, and lack of understanding.
- Broad stakeholder participation is a good indication that ecosystem management is occurring. Often many public involvement meetings are needed as a preliminary to getting anything tangible accomplished.
- Be patient, and keep the big picture in mind! This is a slow and sometimes painful process.

## EXERCISE 2.8

### How to Introduce or Conduct the Exercise

This can be done as three separate groups that then report back to the class, or as one large discussion, facilitated by the instructor.

### What to Expect When Students Do the Exercise

Students may have trouble thinking about any of the organizational types, as they will have little to no experience with organizations. But surely they can identify with bureaucracies, and even pathological organizations, based on their own experiences with obtaining a driver's license, addressing a complaint regarding an improper credit charge, or other common experiences.

### Desired Outcomes for the Exercise

Clear and obvious differences should emerge among the organizational types in how they handle a piece of information. It should be equally as obvious that the adaptive approach is healthier and more effective in addressing problems.

### Points to Be Made Based on the Exercise

- Different types of organizational "cultures" handle information very differently.
- As a consequence, there are major differences in the effectiveness with which they are able to respond to that information.
- Organizations can in fact learn and change, and can move toward the adaptive end of the scale.

## EXERCISE 2.9

### How to Introduce or Conduct the Exercise

See the general description on using the scenario exercises. This is the major exercise for this chapter and is the best opportunity for students to pull together all the information and ideas thus far encountered. The instructor must select four groups of students and assign one task to each, then let them work through it.

### What to Expect When Students Do the Exercise

Students will be confused at first about what their role is to be. Depending on their assignment, they need to either pretend they never heard of ecosystem management and discuss management as though it were the 1960s or 1970s, or fully embrace the ecosystem approach and think in terms of what the chapter has thus far taught them. In either case, it would be most effective if they pictured themselves as members of a resource management agency, such as the USFWS, or whatever is appropriate for that scenario.

### Desired Outcomes for the Exercise

Strong contrasts should be apparent between the traditional and ecosystem approaches for each management situation. Students should see that the traditional approach quickly leads to more tangible actions and decisions made by the agency, often leading to lawsuits. In contrast, the ecosystem approach requires a lot of up-front planning and discussions, bringing people together, building relationships, and giving up and sharing power.

### Points to Be Made Based on the Exercise

Effective ecosystem management requires:

- Much time and effort.
- A willingness to share power.
- Building of trust and relationships.
- Patience and a long-term outlook.

It should result in:

- More effective management over the long term.
- More harmonious human communities.
- Better chances of sustainable resource use.
- Fewer lawsuits and more consensus-based win-win decisions.

## *Teaching Devices to Help Illustrate the Chapter*

- Discussion of how students perceive the role of professionals in natural resource management could help determine their preconceptions and ideas about how management and conservation are done.
- A physical model of the three-context model, cut out of acetate overhead

material and colored in different ways, helps students visualize the three contexts. Discussing them as separate entities and then bringing them together helps illustrate the model and concepts.

## *Sample Test Questions*

1. You are employed in a state or federal resource management agency. Convince a co-worker at your same professional level that moving toward an ecosystem approach is a worthwhile effort.
2. Describe why the command-and-control approach often fails in the long term as a management approach in complex systems.
3. Describe the three basic contexts of ecosystem management and how they interact to lead to effective management.

## *Supplementary Web Sites*

- http://www.snre.umich.edu/ecomgt/ A good general site on ecosystem management from the University of Michigan program, from where Dr. Steve Yaffee has led much of the analyses of ecosystem management efforts. It includes research initiatives, definitions, and a huge number of good links to other sites.
- http://www.great-lakes.net/envt/air-land/ecomanag.html A good example of a regional approach to ecosystem management across political boundaries, with links to many resources related to the subject in and around the Great Lakes. It includes the *Ecosystem Charter for the Great Lakes–St. Lawrence Basin,* a landmark document for the ecosystem approach, both regionally and globally.
- http://ecosystems.fws.gov/ A small site maintained by the USFWS regarding their ecosystem approach.
- http://www.cnie.org/nle/biodv-5.html A Congressional Research Service Report on ecosystem management tools and techniques from a 1995 workshop. It includes the application of various technologies to ecosystem management, along with issues pertinent to the U.S. Congress.

CHAPTER 3

# Incorporating Uncertainty and Complexity into Management

## *Learning Objectives for the Chapter*

THIS CHAPTER BRINGS STUDENTS FACE TO FACE WITH THE REALITIES OF COMPLEXITY and uncertainty in the natural world. Ultimately, this is the basis for nearly everything that follows. Specific learning objectives include:

- Understanding that complexity and uncertainly are underlying features of nature, the human condition, and natural resource management.
- Becoming familiar with various sources of complexity and uncertainty as a way of understanding and dealing with them.
- Learning about general approaches to dealing with complexity and uncertainty in natural resource management.
- Understanding that we need not be paralyzed by or afraid of uncertainty but can use it as an energy source that makes this a fascinating career choice.

## *Major Points of the Chapter*

If students are embarking on a career as natural resource professionals, they need to become comfortable with complexity and uncertainty as driving forces in their lives. If they fight it, they will be frustrated and unsuccessful. If they embrace it as a normal and central part of their careers, they will learn to use it to their advantage, or at least not be paralyzed or intimidated by it.

Complexity and uncertainty can be understood and predicted to some degree. Careful forethought, and decisions made with built-in "buffers" against uncertainty, will help.

## *Items That May Come Up While Teaching This Chapter*

Generally, students understand and identify with the content of this chapter. Not many of them will fight the ideas, and they will seem to embrace them well. They may see some of the "solutions"—such as forecasting events—as unrealistic. Probably the biggest obstacle will be their acceptance of having to provide rather definitive answers in a very uncertain world. This is admittedly unfair, but very much the way it happens—and they should get used to it!

## *Supplementary Activities*

Testimonials from natural resource managers on working with uncertainty can amplify the messages herein. Class participation in a resource management agency's activities also would demonstrate this firsthand.

## *Conducting the Exercises*

### EXERCISE 3.1

#### How to Introduce or Conduct the Exercise

This truly is a "thought exercise." It could be left at that, or you can have class members discuss their thoughts.

#### What to Expect When Students Do the Exercise

Students should develop a heightened sense of awe—not only of nature, but of our limited ability to understand it.

#### Desired Outcomes for the Exercise

An appreciation for the complexities we face and a humility regarding our ability to manipulate and control.

#### Points to Be Made Based on the Exercise

- Nature is hugely complex and fascinating, even at the most limited of scales.
- Humans are naive in thinking they can effectively control large portions of nature over the long term.

## EXERCISE 3.2

### How to Introduce or Conduct the Exercise

This can most effectively be done as a "brainstorming" exercise, where students offer their responses in somewhat rapid fire. Responses are recorded on a blackboard, overhead projector, or flip chart.

### What to Expect When Students Do the Exercise

Students should easily embrace this one and quickly come up with a large list based on their own experiences and ecological knowledge.

### Desired Outcomes for the Exercise

A simple list is sufficient.

### Points to Be Made Based on the Exercise

- There are as many ecological uncertainties out there as our imaginations can invent.
- It should be obvious that these are among many reasons we cannot effectively predict or tightly control ecosystems.

## EXERCISE 3.3

### How to Introduce or Conduct the Exercise

Small groups of the students' own choosing should work well. As with most exercises, getting into groups quickly and getting to work are more important than group composition.

### What to Expect When Students Do the Exercise

This is an opportunity to think in more complex and sophisticated terms than the previous exercise. Students can now express their ecological understanding and make connections that will be helpful in working with their scenario.

### Desired Outcomes for the Exercise

A few good examples from each group are sufficient to demonstrate the point and have students become more involved with the scenario.

### Points to Be Made Based on the Exercise

- Everything is connected to everything else, as we know.
- Pull a string here and something is tugged in an unexpected place. Thus, management must be cautious and flexible.

## EXERCISE 3.4

### How to Introduce or Conduct the Exercise

See the general description on using the scenario exercises. This is the major exercise for this chapter, and it is where students really work to understand complexities and uncertainties in their scenario. We suggest six groups, with two working at each level.

### What to Expect When Students Do the Exercise

Be sure the groups understand their particular focus: (1) all the general ecological things to be thinking about (fire, floods, disease, predation and competition, storms, and so forth); (2) specifically, small populations that are probably endangered or otherwise of concern (and thus demographics and genetics will be relevant, as well as catastrophic events, diseases, and so forth); and (3) the human dimensions—economics, politics (local to national), special interest groups, employment and taxation patterns, institutional players, and so on.

Students should quickly be able to identify several good sources of uncertainty with respect to their focus and begin to think about them in context of the larger system. They also should be able to come up with ways to minimize, head off, or otherwise deal with those uncertainties.

They may try to fill out the list of uncertainties presented in the chapter; this should be discouraged. That information is only background for organizing their thoughts, and it should not be a checklist to be completed.

### Desired Outcomes for the Exercise

After this exercise, students should feel more comfortable about identifying and dealing with sources of uncertainty. They also will be getting more comfortable "inhabiting" their scenario.

### Points to Be Made Based on the Exercise

- Many uncertainties are identifiable and to some degree predictable.
- Forethought and planning pay off.
- Managers need not always be surprised by complexity, and they can prepare themselves for possible negative events. This is akin to members of an emergency management agency: They do not know what type of disaster will occur, when, or where, but they are ready in a general sense because they know that *something* bad will happen and they need to be ready to respond to it.

## *Teaching Devices to Help Illustrate the Chapter*

- Throwing the students an unexpected curve during exercises can help illustrate uncertainties and have them deal with them. For example, change the rules midstream or bring in a new piece of information after they have done a piece of work.
- Computer simulation models that introduce random variation into a system

can help demonstrate the widely disparate results that can be experienced in systems we might otherwise think we can control. This can also demonstrate how models help to assess sensitivities of systems to changes in various parameters.

## *Sample Test Questions*

1. The educated public generally has faith in science and technology to solve most of our problems. Yet, we know that the complexities of natural ecosystems and human behavior prevent science and technology from consistently successful control and sustained production of resources without their degradation. How might you explain this problem to that educated public?
2. Explain how the complexities and uncertainties here feed back to and create the pathology of natural resource management discussed in Chapter 2.
3. Discuss some of the means by which natural resource managers can deal with complexities and uncertainties in their jobs.

## *Supplementary Web Sites*

- http://www.consecol.org *Conservation Ecology*, an online journal, often publishes articles dealing with uncertainty and complexity in natural and managed ecosystems.

# Adaptive Management

## *Learning Objectives for the Chapter*

THIS CHAPTER INTRODUCES ADAPTIVE MANAGEMENT, WHICH IS A KEY CONCEPT FOR the remainder of the book. Specific learning objectives include:

- Understanding the formal definition of adaptive management, as created by its originators.
- Understanding the conditions under which adaptive management can be practiced.
- Distinguishing among three types of adaptive management, described as active adaptive management, passive adaptive management, and documented trial and error.
- Developing examples of these three types for various circumstances.

## *Major Points of the Chapter*

The major point of the chapter is that learning needs to be a fundamental aspect of all management activities and that this is especially true of ecosystem management. Therefore, management actions need to resemble experiments rather than prescriptions.

The chapter adds another dimension to what is typically covered in the literature. We have added "documented trial and error" as a form of adaptive management. The idea here is that even when active or passive adaptive management cannot be performed, management actions can still be thought of as experiments by being explicit about expectations and then collecting data along the way.

A fundamental point of the chapter, and one that runs throughout the book, is that although there is a very formal process called "adaptive management," each of us can try in our own jobs and our own sphere of influence to use the concept to improve our work and our life.

## *Items That May Come Up While Teaching This Chapter*

Several preconceptions can get in the way of understanding this information. First, many people use the term "adaptive management" as any activity in which a person changes through time. It is important to point out that only attempts to learn as you go, rather than just changing in response to current conditions, are really adaptive management.

Second, the formal process of active adaptive management, as defined especially by one of the originators, Carl Walters, is a very elaborate, expensive, demanding process. Consequently, it will seem unrealistic to many readers. In fact, it is unrealistic in many situations, and there are few examples where active adaptive management has been fully implemented. Our premise is that students need to understand the concept in total, and then they can think about how to use it in more practical ways.

Third, younger students especially may have difficulty understanding the difference between how science works and why adaptive management is needed. More elaboration on the scientific method and where it is and is not useful may be required for students to see the limits on learning at an ecosystem scale that are imposed by the strict reductionist nature of scientific experimentation.

## *Supplementary Activities*

The term "adaptive management" is used by most state and federal agencies and some nongovernmental organizations and local conservancies these days. Finding examples from your region (or asking students to find them) will help illustrate how the term is used and how the principles are being used (or not!) in the agencies and groups.

Asking students to think about how the concepts of adaptive management could be used in their student activities, or even the conduct of their education, will help make the ideas relevant to them.

Box 4.1 is a description of the scientific process, but it is very basic. Discussing a research project that you or your colleagues are working on would be much more interesting and informative to students. That discussion can point out the possibilities and limitations of scientific experimentation and get them thinking about the realities of field-based landscape-scale projects.

Similarly, Box 4.2 lists real examples that are considered adaptive management in the implementation of the Northwest Forest Plan. Discussing the characteristics of these examples with regard to the principles of scientific experimentation may help students see the difference between research per se and management-as-experiment.

Box 4.3 lists a variety of conditions needed for adaptive management to succeed. It may be useful to assess a local example of adaptive management (whether called that or not), perhaps even in the university or college, in terms of this checklist so that students can gauge what might or might not work.

## *Conducting the Exercises*

### EXERCISE 4.1

#### How to Introduce or Conduct the Exercise

See the general description on using the scenario exercises. It may be useful to describe the characteristics of biological adaptation and evolution prior to beginning this exercise. This will get the concept into the front of students' minds before having to deal with it in a personal, professional sense.

#### What to Expect When Students Do the Exercise

Younger students may have difficulty thinking about how a nonadaptive person will react. The easiest example for students to work with here is via SnowPACT, in which both approaches might lie within the experience of a student (fighting or accepting a law/rule change).

#### Desired Outcomes for the Exercise

The desired outcome is for students to see how different the approaches are when people try to adapt or not. They should see two very different personalities.

#### Points to Be Made Based on the Exercise

- Approaches can be very different.
- How you approach something, in your head and via your behavior, matters.
- You have a choice.
- An adaptive person is open-minded and collaborative.
- Being adaptive is necessary in conditions where change is likely and the environment is unstable—hence, in modern life and ecosystem management.
- Diversity in people and organizations is both useful and desirable because it enhances adaptiveness.

## EXERCISE 4.2

### How to Introduce or Conduct the Exercise

This exercise needs to be done in groups, because individuals will have trouble thinking of items to include in any cell of the table. We recommend reviewing the first two rows of the table before putting students to work.

### What to Expect When Students Do the Exercise

Students may have trouble noting problems with scientific experimentation, having been educated to be scientists. Therefore, some initial work to help them see that science isn't always the answer may be needed.

### Desired Outcomes for the Exercise

The desired outcome is the recognition that science is good in simple reducible situations; that tradition is useful in stable situations; that the trial-and-error approach is hard to transfer or apply broadly.

### Points to Be Made Based on the Exercise

- The best ways to learn and know vary depending on the situation.
- Nothing is really very good for the large-scale situations in ecosystem management.
- Therefore, we need to be open to new ways to learn.

## EXERCISE 4.3

### How to Introduce or Conduct the Exercise

See the general description about using the scenario exercises. This is the major exercise of the chapter. It will probably need to be conducted over a substantial time (1 week) or during an extended discussion or laboratory session. The PDQ Revival topic is the simplest; the ROLE Model topic is the easiest to think of in terms of adaptive management; the SnowPACT topic provides the most opportunity for innovative thought.

### What to Expect When Students Do the Exercise

The biggest obstacle in this exercise will be defining the difference between active and passive adaptive management at each site, because the distinction between the styles would be clearest with two entirely different sites.

### Desired Outcomes for the Exercise

The specifics are much less important than the main ideas, which are that these circumstances should be management-based (not research projects) and that distinctions are made from active to passive to documented trial and error.

**Points to Be Made Based on the Exercise**

- Real differences exist among the approaches.
- Each approach may be more useful in various circumstances (i.e., documented trial and error for nest boxes; passive adaptive management for dam removal [the choice of which dam to remove may depend on other circumstances]; active adaptive management for bison introduction).
- Discussion of the conditions necessary to achieve each approach would be useful.

## *Sample Test Questions*

1. Describe the characteristics under which an adaptive management approach would be most and least desirable.
2. Describe the argument you would make to convince a resource manager that the next action taken should be documented trial and error.
3. Given this circumstance (create a circumstance based on a local or state resource issue), what kind of adaptive management approach would you recommend, and why?

## *Supplementary Web Sites*

- http://www.consecol.org The online journal *Conservation Ecology* publishes many articles on adaptive management.
- http://www.fws.gov/r9mbmo/mgmt/ahm2.html The U.S. Fish and Wildlife Service's waterfowl management program has adaptive information.
- http://www.for.gov.bc.ca/hfp/amhome Government of British Columbia keeps an adaptive management Web site.
- http://www.uc.usbr.gov/amp/ The Glen Canyon Dam Adaptive Management Program has an extensive Web site.
- http://www.or.blm.gov/roseburg/littleriverama/index.htm The Little River Adaptive Management Area is a good link to the adaptive management program in the Pacific Northwest.

# Genetic Diversity in Ecosystem Management

## *Learning Objectives for the Chapter*

THIS CHAPTER BEGINS THE TECHNICAL SECTION OF THE BOOK. IT PROVIDES A PRIMER IN conservation genetics, and subsequent chapters move up through populations, species, and the landscape. Specific learning objectives include:

- Understanding basic conservation genetics concepts and terminology.
- Understanding the role of heterozygosity in fitness.
- Becoming familiar with the concept of genetically effective population size.
- Understanding how quantitative genetic variation and allelic diversity are lost in small populations.
- Understanding the geographic distribution of genetic variation and why that can be an important resource to protect.
- Tying all of the above back to management decisions in a larger ecosystem context.
- Understanding the limitations of genetics in conservation and ecosystem management.

## *Major Points of the Chapter*

The chapter begins the foray into technical content and tries to give students the basics of conservation genetics. It is not meant to be comprehensive (there are entire texts and courses for that) but rather to give students a working understanding of the genetic level of biological organization.

Genetic diversity is a resource that can be degraded just as any other resource. It can be lost in small populations and by artificial gene flow among different populations, and these losses can have real fitness consequences. The students are presented with a few relatively simple quantitative exercises to get a feel for how this works.

Genetics is a tool, and like any tool, it has limited capabilities. Students must realize that it is very important for some uses but inappropriate for others.

## *Items That May Come Up While Teaching This Chapter*

Some students will fear the technical or mathematical aspects of genetics. This is not usually an actual problem, as the material is explained carefully and slowly enough for even very inexperienced people to grasp it.

Students may want to discuss famous cases where genetics has been used or has been a concern—African cheetahs, the Florida panther, captive rearing, aquaculture, hybrids, and so forth. This is a great opportunity to tie the theory back to real applications, and we suggest you accommodate these discussions as much as possible.

Managers often want to know specific answers to their questions regarding management issues with a genetic basis. Typically, there are no straightforward answers because every situation is different. Students need to realize this. There is also a lot of room for values here: What exactly are we trying to preserve? For example, do we want Florida panthers or panthers in Florida? The answer may drive how the situation is treated genetically.

## *Supplementary Activities*

The CD-ROM *Conserving Earth's Biodiversity* by E.O. Wilson and Daniel L. Pearlman has a genetic stochasticity model that can show gene frequency changes as a function of different population sizes. The effective population size model shows how sex ratios affect $N_e$.

A guest appearance by a population geneticist or knowledgeable hatchery manager who has used genetics could provide a productive interchange.

## *Conducting the Exercises*

### EXERCISE 5.1

**How to Introduce or Conduct the Exercise**

This is a thought exercise and could be left at that. Or, you could have a brief discussion to see what the students came up with.

### What to Expect When Students Do the Exercise

If a discussion is pursued, students should be able to suggest a number of measures that could be used, such as number and size of offspring, disease resistance, number of matings garnered, relative growth rates, or measures of asymmetry.

### Desired Outcomes for the Exercise

Several other measurements should suffice.

### Points to Be Made Based on the Exercise

Many fitness-related traits could be tied back to genetic diversity.

## EXERCISE 5.2

### How to Introduce or Conduct the Exercise

This is best conducted as an open class discussion.

### What to Expect When Students Do the Exercise

Students should be able to freely exchange ideas at this point and have enough information to make reasoned deductions.

### Desired Outcomes for the Exercise

An understanding that management decisions can and should be made with genetics in mind where appropriate. Relatively simple decisions based on even minimal information can have important consequences for future genetic constitution.

### Points to Be Made Based on the Exercise

- Maximizing the number of males and females in breeding programs will maximize the preservation of genetic diversity.
- Having mass production—if a result of one or a few individuals of one sex—is not desirable. It would be better to have fewer offspring if they represent a wider variety of parental genes than many offspring from a few parents.

## EXERCISE 5.3

### How to Introduce or Conduct the Exercise

This is best conducted as an open class discussion.

### What to Expect When Students Do the Exercise

Again, students should have little problem in exchanging ideas for this exercise.

### Desired Outcomes for the Exercise

This is a good opportunity to tie a theoretical aspect of genetics directly to a real management action. Students should be able to see the practical application of even a rudimentary understanding of genetics.

### Points to Be Made Based on the Exercise

- The founding stock defines the future of that reintroduced population and is thus important to think about.
- The size and sex ratio of the founders determine the $N_e$ for that stock and can be manipulated accordingly.
- Genetic variation of the reintroduced population can best be maximized with a larger number of founders and an even sex ratio.

## EXERCISE 5.4

### How to Introduce or Conduct the Exercise

It is best to assign these numbers to small groups of students working together; do not have every student calculate every $N_e$ or it will take too long. Draw a graph with the $N_e$s across the bottom and 0 through 100% on the y-axis. Plot the various points the students calculate, and connect them. You should have an asymptotic line starting at 50% and approaching 100%.

### What to Expect When Students Do the Exercise

This should be an easy calculation for most of the students; they are simply plugging their number into the formula $1 - [1/2N_e]$.

### Desired Outcomes for the Exercise

A smooth curve and an understanding of the behavior of $N_e$ at small population sizes is what we are after.

### Points to Be Made Based on the Exercise

- Small populations lose genetic variation more quickly than large populations.
- From a management perspective, significant genetic variation is not lost in one generation unless $N_e$ gets below about 10 or 20.

## EXERCISE 5.5

### How to Introduce or Conduct the Exercise

This is one of the two major exercises for the chapter, and it gets the students working with most of the genetic concepts learned to this point. Again, students should work in small convenient groups.

### What to Expect When Students Do the Exercise

Students should be able to deal with the relatively simple math involved, as conceptual understanding of the issues is really the driving force; the math is trivial.

### Desired Outcomes for the Exercise

The following numbers are what they should be arriving at:

Original $N_c$: 100 Original $N_e$: 100
New $N_c$: 50 New $N_e$: 32 (calculated as $4(40 \bullet 10)/(40 + 10) = 1600/50 = 32$)
$N_c$ declined by 50 $N_e$ declined by 68
Proportion of genetic variation retained: 98.4% (calculated as $1 - 1/2N_e = 1 - 1/64$)

Management recommendations you receive might include the following, listed in increasing order of intensity and effort: don't do much and monitor the population to see what happens; try to find out why males differentially disappeared (did they die or leave the population?); do some basic genetic analyses to see what we are working with; start up a new population as a hedge against further losses (where should these individuals come from?); start a captive breeding program to increase the population size.

There is a tendency for students to want to charge out and do some huge studies, analyses, or major management actions. We suggest management caution and using this information only as an alert that something may (or may not) be wrong.

### Points to Be Made Based on the Exercise

- Basic demographic information can be used to alert the manager to possible genetic problems.
- Don't overreact! Use this information within a larger context. What is really going on here? Do we have a genetic crisis (probably not) or something of demographic concern (more likely)? It is tempting to take genetic knowledge and run wild with it. Instead, use it as part of a larger toolbox.

## EXERCISE 5.6

### How to Introduce or Conduct the Exercise

This is the second major exercise for the chapter, and it is quite challenging. It may be best to do this as a class exercise and work it through as a group. (Alternatively, students may form groups of five or six.)

### What to Expect When Students Do the Exercise

Students will really stretch themselves and be forced to use their knowledge of genetics in the context of broader skills. They definitely will need a calculator.

### Desired Outcomes for the Exercise

This is really an exercise in genetic drift, using the equation $[1 - (1/2N_e)]^t$. We want that number to be at least 95% (i.e., a loss of no more than 5%). The key here is t, or number of generations. We are told that generation time is 6 years, and we are concerned about genetic losses over 100 years, so how many generations is that? The answer is 100/6 = 16.67, or 17 generations. So we know that $[1 - (1/2N_e)]^{17} = 0.95$. Stated another way, $X^{17} = 0.95$, where X is $1 - (1/2N_e)$. X, then, should be the 17th root of 0.95, which on the calculator equals 0.99698. Therefore, $1 - (1/2N_e) = 0.99698$, which means $1/2N_e = 1 - 0.99698 = 0.00302$. Inverting, $2N_e = 331$, and $N_e = 165.56$ or 166. Because the $N_e/N_c$ ratio is 0.2, the census number must be five times $N_e$, or 830. If 25,000 acres are needed per bear, this comes to over 20 million acres.

### Points to Be Made Based on the Exercise

- Genetic information can be used to estimate management requirements for species, even on a huge landscape scale.
- Even problems that seem imposing can be worked through with just a little bit of logical thought.

## EXERCISE 5.7

### How to Introduce or Conduct the Exercise

This will work best in small groups of four or five individuals. Students should conduct a general, open discussion, and then share their conclusions with the class.

### What to Expect When Students Do the Exercise

Students should come up with a fairly wide ranging set of ideas, the specifics of which depend on which species they choose.

### Desired Outcomes for the Exercise

Students should explore the ramifications of gene flow among populations of a species within the scenario. They should realize that it is important for managers to understand historic patterns of gene flow and geographic relationships of the landscape before contemplating genetic manipulations among populations.

### Points to Be Made Based on the Exercise

- Populations that have been isolated for a long time may have evolved adaptations to local conditions, and such adaptations could be jeopardized by artificial gene flow.
- Conversely, populations that have historically experienced regular gene flow could be jeopardized by human-induced isolation.

- Distinguishing these two situations is fundamental to decisions regarding species manipulations and gene flow.

## EXERCISE 5.8

**How to Introduce or Conduct the Exercise**

This should be a general group discussion. Remind the class about the conclusions from Exercise 5.5, and then pose the question.

**What to Expect When Students Do the Exercise**

This should be a rather quick exercise with few surprises.

**Desired Outcomes for the Exercise**

Students should be able to see the difference between losing quantitative genetic variation and losing specific, rare alleles.

**Points to Be Made Based on the Exercise**

- Even when quantitative variation is lost slowly, it is possible for real losses of rare alleles to occur even in one generation.
- Losses of rare alleles could have immediate fitness consequences for the population, probably more so than for losses of quantitative variation.

## *Teaching Devices to Help Illustrate the Chapter*

- The new U.S. quarter helps illustrate the idea of multiple alleles at a gene locus. There are two "loci" for all U.S. quarters—heads and tails. The head has only one possible allele: George Washington. But the new U.S. quarters have many different possible "tail alleles"—one for each state. Each one is the same locus, a tail, but the expression of that locus varies widely.
- Well-known examples of conservation genetics problems from the literature such as the Florida panther and the African cheetah are helpful.
- The loss of crop genetic diversity, the narrow genetic base of many food crops, and the possible ramifications for humanity's food supply call attention to the potential dangers of genetic losses.
- Inbreeding in closed human societies helps illustrate inbreeding depression in a clear way.

## *Sample Test Questions*

1. Describe the relationships among founder effect, genetic bottlenecks, genetic drift, and inbreeding depression. How are they similar? How do they differ?
2. Why is it important from a management perspective to know how total

genetic diversity is partitioned into within- and among-population components?

3. Discuss some of the necessary genetic considerations involved in translocating populations for the following scenarios: (a) into an area where the species was native but no longer occurs; (b) into an area where the species occurs but is on the decline.

## *Supplementary Web Sites*

- http://www.consbio.umn.edu/ConsGen/ This University of Minnesota Web site lists journals that publish conservation genetics materials, good review articles and books on the subject, available software, and major researchers in the field listed by primary taxa and topics they pursue, all with links to more information.
- http://gslc.genetics.utah.edu/society/conservation/ A fairly elementary but useful site from the University of Utah that describes conservation genetics, including its tools and uses, and provides some simple problems.
- http://www.rom.on.ca/biodiversity/cbcb/cbmolecu.html A site from the Royal Ontario Museum showing various ways in which conservation genetics is used in its research programs.

CHAPTER 6

# Issues Regarding Populations and Species

## *Learning Objectives for the Chapter*

THIS CHAPTER CONTINUES UP THE SCALE OF BIOLOGICAL COMPLEXITY FROM GENES TO populations and species. Specific learning objectives include:

- Understanding that "species" is only one way to look at biological diversity, and that it masks a great deal of variation.
- Making a connection between genes (previous chapter) and landscapes (following chapters).

## *Major Points of the Chapter*

The "species" is a flexible and changing entity, sometimes ill-defined, and subject to change. Regardless of definitions used, a great deal of populational variation lies within the species. Populational variation is important for conservation because of the genetic variation populations carry, the local adaptations they may have, and the functions they perform in their ecosystems. Species can play a variety of roles in policy making and public perception, depending on how they are viewed. We present six ways to look at species relative to policy and perception. Populations and species occur on landscapes in particular ways. We begin to address this here (to be followed up in later chapters) by discussing alpha, beta, and gamma richness.

## *Items That May Come Up While Teaching This Chapter*

If students have a favorite species concept, they may try to defend it. But that is not the point here; rather, they need to understand the great diversity that resides within and is hidden by "the species."

There is a tendency to call anything a "keystone species" if it affects any other species in the system. Stick to the definition of keystone as having influences out of proportion to its biomass. Keystones are also to be distinguished from structural species, such as trees, which physically dominate a system but are not necessarily keystones.

## *Supplementary Activities*

The CD-ROM *Conserving Earth's Biodiversity* by E.O. Wilson and Daniel L. Pearlman has a good module called "Diversity of Life" that would supplement some of this chapter's material. It includes the relative numbers of different known taxa and the estimated numbers yet to be discovered.

## *Conducting the Exercises*

### EXERCISE 6.1

#### How to Introduce or Conduct the Exercise

The instructor will need to lead this class discussion.

#### What to Expect When Students Do the Exercise

Students may not know many definitions, at least not formally. You may need to have some prepared materials that compare the biological, phylogenetic, and a few other major species concepts. But we suggest keeping it simple and sticking to the main point: There are different definitions being used, and they result in different views of species.

#### Desired Outcomes for the Exercise

A greater familiarity with various species concepts by the students.

#### Points to Be Made Based on the Exercise

- There are various species definitions in use, with no universal acceptance.
- The different definitions can affect the legal protection of species, depending on how the "species" is viewed.

## EXERCISE 6.2

### How to Introduce or Conduct the Exercise

Again, the instructor should lead this quick class discussion.

### What to Expect When Students Do the Exercise

Students should have a basic understanding of natural selection. If they are rusty, some review may be needed.

### Desired Outcomes for the Exercise

A good understanding of the mechanics of natural selection.

### Points to Be Made Based on the Exercise

- Genetic variation is generated randomly.
- Selection is not random.
- For evolution to operate, the selected traits must have a genetic basis and be inherited.

## EXERCISE 6.3

### How to Introduce or Conduct the Exercise

Get the students into small groups of four or five, and have them quickly focus on one species from their scenario.

### What to Expect When Students Do the Exercise

Students should quickly be able to address the various points using their species.

### Desired Outcomes for the Exercise

Students should come up with several ramifications of the species' loss—genetic, adaptations, and functions. Repercussions for other species may be more difficult to come by, and more speculative. This is a good opportunity for them to begin thinking about their biological conclusions in a social setting and figuring out ways to make their conclusions understandable to a diverse set of stakeholders.

### Points to Be Made Based on the Exercise

- Populations lost from an ecosystem can have serious repercussions for that system, regardless of their continued existence elsewhere.
- Endangerment is only one criterion by which species may be judged, and populational losses may be significant.
- Making such information relevant to stakeholders is not necessarily easy, and it requires skills that go beyond biological knowledge.

## EXERCISE 6.4

### How to Introduce or Conduct the Exercise

Decide whether you would prefer to do this as a class discussion or in small breakout groups.

### What to Expect When Students Do the Exercise

Students should be somewhat challenged in discussing some of these questions and may need to think about them deeply.

### Desired Outcomes for the Exercise

A better familiarity with these issues based on having to think through some of these points.

### Points to Be Made Based on the Exercise

- Thinking about vulnerability, keystones, umbrellas, and the like helps us understand their roles in the ecosystem and how they might be presented to various stakeholders.

## EXERCISE 6.5

### How to Introduce or Conduct the Exercise

It might be best for the instructor to select the species. Focus on those species students are likely to know something about, so they can make reasoned judgments. You may wish to prepare a one-page matrix in advance with the species listed along the left side, and then six columns for the species categories, which may be checked off as appropriate.

### What to Expect When Students Do the Exercise

You can expect both strong congruence in pattern for some species and categories, and a scattering of responses for others. It is the scattering that is interesting and worth pursuing in a discussion, asking why students think the way they do about some of these categorizations.

### Desired Outcomes for the Exercise

A completed chart that reflects every student's input, followed by a class discussion about the patterns that developed. You should pursue a discussion about specific points—for example, why only 2 out of 28 thought a given species was a keystone. What does this mean for interactions with the general public?

### Points to Be Made Based on the Exercise

- A given species may play multiple ecological and political roles in a system; other species may rarely show up on the charts. (Does this mean the latter can be ignored?)
- Often there is little congruence for how a given species might be judged, even by a class of students with biological training. This should speak to the challenges and opportunities of dealing with various stakeholders regarding the species in question.

## EXERCISE 6.6

### How to Introduce or Conduct the Exercise

This exercise is probably best conducted as an open class discussion.

### What to Expect When Students Do the Exercise

Students will have to contemplate these questions just a bit and think through the ramifications.

### Desired Outcomes for the Exercise

By this point students should begin to think about species in a landscape context and recognize that management manipulations of habitats can greatly affect species composition and richness at the landscape level. This is not always easy or straightforward. Also, some value issues arise, regarding desired conditions. By whom? And for what?

### Points to Be Made Based on the Exercise

- Alpha, beta, and even gamma richness may be manipulated by management decisions. Be careful!
- Manipulation for one type of species, or richness at one level, can have unintended effects at other levels.
- These are yet more reasons why management must be conducted across political boundaries and with input from many stakeholders.

## EXERCISE 6.7

### How to Introduce or Conduct the Exercise

We suggest splitting the class into three or six groups, depending on how many students you have. With six groups, you have replicates. Each group should work on one of the issues and then report back to the class.

### What to Expect When Students Do the Exercise

Students probably will struggle just a bit but then put together the pieces necessary to move forward.

### Desired Outcomes for the Exercise

We seek innovative thoughts on the part of the students that demonstrate a good understanding of how management actions affect alpha, beta, and gamma richness. Good discussions should ensue.

### Points to Be Made Based on the Exercise

- Manipulation of one aspect of richness usually affects others. Thus, be careful when manipulating habitats!
- Management at larger scales can significantly affect how species are distributed across landscapes, and thus patterns of species richness.

## *Sample Test Questions*

1. Why is the "species" a hypothesis rather than a solid, unchanging construct?
2. Why is within-species variation considered so important in conservation?
3. Discuss the importance of seeing species in a biopolitical sense when interacting with stakeholders.
4. Why are alpha, beta, and gamma richness important concepts relative to management decisions?

CHAPTER 7

# Populations and Communities at the Landscape Level

## *Learning Objectives for the Chapter*

THIS CHAPTER PRESENTS AN OVERVIEW OF MANAGING FOR BIODIVERSITY AT THE SPECIES level and at the level of species communities (collections of species). Note that the single-species approach focuses on species of conservation concern—those that are threatened and experiencing population declines. Specific learning objectives include:

- Understanding that a declining species has both a proximate factor and an ultimate factor that led to its decline.
- Understanding the difference between deterministic and stochastic forces that lead to population declines.
- Appreciating the three general approaches used to estimate a population size that would be able to persist over time: minimum viable populations (MVP), metapopulations, and spatially explicit models. These three methods operate at different spatial scales and require different levels of data sophistication, beginning with the most simple (minimum viable populations) and moving to the most complex (spatially explicit models).
- Understanding that managing biodiversity one species at a time has serious difficulties. A complementary approach is to also manage for collections of species.
- Being able to explain three different approaches that can be used to manage for collections of species. Also being able to explain ways to monitor the effectiveness of each approach and being aware that all three are often

used collectively at a landscape level to ensure the persistence of native biodiversity.

## Major Points of the Chapter

Ecosystem management differs from traditional management in that it is concerned with all native species, not just those that have economic value (Douglas fir trees, white-tailed deer), those that threaten our economies (prairie dogs, red-winged blackbirds), or those that are endangered. To ensure the maintenance of native biodiversity, managers need to employ a variety of methods, focusing on both single species and collections of species, and at a variety of spatial scales. This requires various approaches, ranging from modeling to restoring natural ecological processes on landscapes. These methods also require a variety of data needs, some relatively simple, others much more detailed. Students need to become adept at these various approaches and be able to understand how to monitor them to ensure they are working. Students also need to become comfortable with the concept of using models to either guide management actions or enable them to ask questions that would be impossible to address empirically at the landscape level.

## Items That May Come Up While Teaching This Chapter

Students may not at first understand the difference between proximate and ultimate factors that lead to a population decline. It is important to emphasize both factors and ensure that students can distinguish them. Using species they are familiar with and discussing proximate and ultimate factors in the species' declines will facilitate understanding.

It is equally important that students see the connection between deterministic and stochastic forces that lead to a population decline. Deterministic forces operate at the landscape level to decrease and isolate populations. Depending on the size of these populations, they become increasingly at risk because of stochastic forces. Here students should refamiliarize themselves with the information presented in Chapter 3. By using examples, they will come to understand that deterministic forces set up populations that have become isolated and reduced in number to the effects of stochastic forces.

When students understand the amount of detailed demographic and landscape-level data needed to model species populations, they may very well throw up their hands in despair. After all, very few species have been studied in such detail. This can be a good "teachable moment" by emphasizing that PVA, metapopulation, and spatially explicit model approaches "should not be tried at home." That is, they should be undertaken with the utmost seriousness, because the results from these modeling efforts often have profound implications in guiding management actions. On the other hand, models should not be disparaged because they can have great heuristic power in allowing managers to ask "what if" questions they would never able to ask at the landscape level. In other words, these three modeling approaches can be profoundly useful, but their strengths and limitations have to be acknowledged.

Students may not appreciate that most efforts at estimating minimum viable population sizes use the modeling approach. It is important to emphasize that the two other approaches—observational and experimental—should be considered as well. Land management and natural resource agencies have largely ignored the opportunity in ongoing land-management activities (such as logging or grazing) to use these two approaches. This fact can be used to emphasize how important it is to employ adaptive management as a way of doing business.

Introducing MVP, metapopulations, and spatially explicit models is a great opportunity to emphasize how our thinking is becoming more spatially realistic relative to managing species. Modeling MVPs is done for a spatially discrete population; metapopulations acknowledge multiple, spatially discrete populations connected via dispersal (but they say nothing about the landscape on which dispersal occurs); and spatially explicit models include probabilities of survival and reproduction at a landscape level for multiple populations.

When teaching concepts associated with managing for species communities, a variety of thoughts may arise. With few exceptions, most management has tended to focus on single species; the idea of managing for a collection of species may very well come across as vague in the eyes of students. This is understandable because the two approaches are quite different. Students may be able to easily understand managing species because efforts tend to focus on birth, death, and dispersal rates. But when managing for collections of species, students may not be as comfortable thinking about ecological processes, landscape elements, or selected species that affect many.

Students also may have difficulty imagining how these three approaches can be monitored for effectiveness. After all, these methods seldom require counting a species' population. Instead, they may appear a bit amorphous because they tend to measure ecological processes or changes in landscape elements. These sharp contrasts are great opportunities to emphasize how ecosystem management has contributed to the ongoing evolution of natural resource management.

## *Supplementary Activities*

Students should be encouraged to visit a variety of Web sites where they can actually simulate species population changes over time using PVA, metapopulation, or spatially explicit models.

Guest speakers from natural resource management agencies could be useful in highlighting points made in this chapter, particularly if they are managing landscapes utilizing the newer ideas under both single-species and multiple-species approaches. Videos might also be useful because they can show students how agencies are trying to reinstate fire, grazing, and hydrological regimes to landscapes that have long been denied these natural ecological processes.

## *Conducting the Exercises*

### EXERCISE 7.1

#### How to Introduce or Conduct the Exercise

This exercise uses the scenario the students are working on. They need to form small groups and use the species that are declining and that are of conservation concern. Give them 10–15 minutes to develop their responses, then have each group's spokesperson present the group's answer to the class.

#### What to Expect When Students Do the Exercise

The students will initially struggle with the difference between proximate and ultimate factors that lead to a population decline. Have them revisit the material in the textbook as well as look up words in a dictionary.

#### Desired Outcomes for the Exercise

They should be able to list at least one proximate and one ultimate factor for the decline of at least one species in the scenario. If they can do more, so much the better. The bottom line is for them to understand the difference between the two and be able to recognize them for other species.

#### Points to Be Made Based on the Exercise

- Proximate factors in a species' decline are usually aspects of birth and/or death rates.
- Ultimate factors in a species' decline are usually aspects of the species' environment, such as the conversion of ecosystems to some human use, or a widespread use of a toxin, such as a pesticide.
- Ultimate factors always precede the proximate factors and, indeed, drive the proximate factors (clear-cutting forests will lead to a decreased birth rate and an elevated death rate of a forest-dependent species).

### EXERCISE 7.2

#### How to Introduce or Conduct the Exercise

Break the class into small groups and have students consult the scenario they are using. Have them pick two species and for each one design a study that could determine an MVP for that species. Have them develop an observational approach for one species and an experimental approach for the other species. Give them 15 minutes, then have each group's spokesperson present the group's ideas to the class.

### What to Expect When Students Do the Exercise

Some groups may have difficulty distinguishing between the observational and the experimental approaches. Have them revisit the text and discuss the examples used to illustrate the two approaches.

### Desired Outcomes for the Exercise

Each group should be able to clearly distinguish the two approaches and illustrate the differences by the species they choose to use.

### Points to Be Made Based on the Exercise

- The observational approach to estimating a species' MVP is based on monitoring populations of different sizes; these different-sized populations usually occur in areas of different sizes (i.e., larger areas have larger populations).
- The experimental approach requires manipulating landscapes so that different-sized patches occur, each supporting different-sized populations.
- Both approaches require long-term monitoring of population sizes in order to determine the trajectory of the populations.
- Both approaches assume that the patches that support the populations are similar. This is, of course, an unrealistic assumption, particularly for the observational approach.

## EXERCISE 7.3

### How to Introduce or Conduct the Exercise

The instructor should lead this class discussion after the students are aware of the differences between the deterministic and stochastic modeling approaches to estimate an MVP.

### What to Expect When Students Do the Exercise

Some students are not likely to understand the differences in the two approaches. This is a great opportunity to emphasize that the deterministic approach uses averages for fecundity and survival, whereas the stochastic approach allows the variance in the data to express itself. Also, few students will know why there are seldom differences in the outcomes of the two approaches. This will allow you to point out that with only 4 years of data, the chance of much variation in fecundity and mortality to occur is slight. It would take many years of data to actually represent environmental and other kinds of variation discussed in Chapter 3.

### Desired Outcomes for the Exercise

At the end of the discussion, students should be aware of the differences between the two approaches. They should also understand that there may be few differences between the two methods if the demographic data used in the models fail to capture stochastic variation that may occur over a long period of time.

### Points to Be Made Based on the Exercise

- Deterministic models use the averages of demographic data, whereas stochastic models use the variation inherent in any data set.
- Deterministic models will give you the same answer repeatedly unless you change the demographic data or the number of individuals in each age class. Stochastic models will give you different answers because they use different numbers every time they are calculated.
- Unless the data used have been collected over an extended period of time, and therefore represent variation in birth and death rates, they may give very similar answers using either method.

## EXERCISE 7.4

### How to Introduce or Conduct the Exercise

The instructor should lead this discussion. Make sure a species is chosen that occurs in the scenario the class is working on and that occurs in spatially discrete populations (a metapopulation).

### What to Expect When Students Do the Exercise

This exercise is a good opportunity to ensure students understand the concept of metapopulations and its associated terms. Also, it will enable the students to come to grips with the difficulties of managing populations that are spatially discrete and occur on lands with different ownership.

### Desired Outcomes for the Exercise

Students should discuss the dilemmas associated with a population that is a sink (i.e., what to do with it). Also, they should recognize the challenges associated with managing a species that occurs in an ecosystem fragmented by diverse administrative boundaries.

### Points to Be Made Based on the Exercise

- Populations often consist of spatially discrete populations that occur on lands with diverse ownerships.
- Discrete populations may exist as sources or sinks, and the viability of a metapopulation may depend on how many of each there are and whether dispersal can occur among them.
- Discussions on whether to manage for the persistence of spatially discrete populations may depend on whether they are sources or sinks, as much as whether they occur on protected areas or on privately owned land.

## EXERCISE 7.5

### How to Introduce or Conduct the Exercise

Break the class into groups. For their scenario, have each group prepare a management plan for a species that exists as spatially discrete populations. Each group should take an MVP approach, a metapopulation approach, or a spatially explicit approach. Afterward, have the groups participate in a three-way discussion on the strengths and weaknesses of the three approaches.

### What to Expect When Students Do the Exercise

Students should have an appreciation of how each of the three approaches works, how they are different from each other, and how challenging it would be to develop management plans for any of the three approaches.

### Desired Outcomes for the Exercise

All groups should list at least one management idea for their assigned approach. They should be aware of assumptions they had to make in order to employ their approach. They should also be aware that, based on the land ownership where the species occurs, their management steps may differ (e.g., using the MVP approach, they may do one thing for a population on private land and something quite different if it occurs on public land).

### Points to Be Made Based on the Exercise

- Each of the three approaches requires different levels of information, with the spatially explicit approach most information demanding. None of the three is easy!
- Each of the three methods is best done using an adaptive management approach; certainty is a luxury!

## EXERCISE 7.6

### How to Introduce or Conduct the Exercise

Break the class into small groups, and assign each group one of the three approaches (species, ecological process, or landscape) to managing for a group of species in the scenario they are working with. Each group should report back to the class, which then should prepare a list of the strengths and difficulties of the three approaches.

### What to Expect When Students Do the Exercise

Students should be able to explain each of the three approaches and be able to distinguish each from the others.

### Desired Solution for the Exercise

A list of strengths and difficulties associated with each of the approaches should result as well as a list of information needs for using each of the three methods.

### Points to Be Made Based on the Exercise

- The species approach focuses on a species that affects many other species. Students may confuse this approach with the traditional single-species approach in natural resource management.
- The ecological process approach requires knowing the spatial and temporal scales within which the process naturally occurs. This may not be already known.
- The landscape approach will require using GIS technology and being aware of size, shape, area, and composition of landscape elements, such as old-growth forests, wetlands, and human uses of the area.
- These approaches are seldom done in isolation. They are best done as an integrated management approach.

## *Teaching Devices to Help Illustrate the Chapter*

- Have students consult newspaper stories and magazine articles that detail the declines of threatened or endangered species. From these accounts, have them distinguish the proximate and ultimate factors leading to the decline. How often are these factors not human-caused?
- Have students contact the state wildlife agency and find out for what species they have long-term trend data for different-sized populations. Perhaps personnel from the agency would be willing to discuss their approach to ensuring viable populations of certain species in the state.
- Students could consult various scientific journals, such as *Conservation Biology*, and write brief reports on papers that use an MVP, metapopulation, or spatially explicit approach.
- Computer models that let students calculate an MVP of a species using the stochastic and deterministic approaches will allow them to see how the numbers change, or do not change, with different numbers of iterations.
- Students could be asked to catalogue the number of papers in appropriate conservation science journals and list how many papers focus on single-species management versus species community management. For both approaches, are certain taxa more reported on than others?

## *Sample Test Questions*

1. You have just conducted a PVA for a rare species of orchid. Your lambda value is 0.996. What does this suggest about the population? You may recall that you used the modeling approach to derive this estimate. Describe two other approaches one could use to estimate an MVP.
2. Prairie dogs are being considered for protection under the Endangered Species Act. They exist in towns that are spatially discrete. Prairie dogs are

also good dispersers. What characteristics of the populations do you need to measure to determine whether the populations exist as a metapopulation? How would you determine whether a particular town is a source or sink? If all the sources are found on private lands and the sinks are on public lands, what implications would this have for the conservation of prairie dogs? What are the conservation implications if all the sources are found on public lands?

3. Define a metapopulation. When prioritizing protection, which populations of a metapopulation should a natural resource manager focus on? Why? Which populations are less important when prioritizing efforts? Why? Knowledge of metapopulations has enabled managers to develop more specific guidelines for designing nature reserves. Describe three of these guidelines, and relate them to the concept of metapopulations.
4. Most management for biodiversity is being done one species at a time. List four challenges with managing this way. List and explain three approaches one can use to manage for species assemblages. Suggest one way you could monitor the success of each approach.

## *Supplementary Web Sites*

- http://www.consecol.org *Conservation Ecology*, an online journal, publishes articles utilizing MVP, metapopulation, and spatially explicit models.
- http://www.conserveonline.org This site, produced by The Nature Conservancy, is a collection of practical information for students interested in conservation, including journal articles, unpublished reports, case studies, presentations, maps, and software. The documents can be downloaded or viewed online. Students can rank resources and write reviews.
- http://www.fs.fed.us/library/ This U.S. Forest Service library of publications and databases is available to anyone. Although you cannot search by keyword, the database is categorized by type of publication or data set. There are data sets pertaining to, for example, fire management, climate change, and regional flora—as well as maps and Forest Service software for downloading.

CHAPTER 8

# Landscape-Level Conservation

## *Learning Objectives for the Chapter*

THIS CHAPTER ADDRESSES BIODIVERSITY AT A LANDSCAPE SCALE AND INCLUDES THE ramifications of fragmenting ecosystems. Specific learning objectives include:

- Understanding that fragmenting an ecosystem reduces the area, increases the amount of edge, and increases the isolation of ecosystem remnants.
- Appreciating that certain species are influenced by these three effects of fragmenting ecosystems. Reduced area affects area-sensitive species, increased edge affects edge-sensitive species, and increased isolation of ecosystem remnants affects dispersal-sensitive species.
- Understanding that the depth of edge effects varies depending on whether the edges are biotic, abiotic, or human-caused.
- Appreciating that the shape and area of the ecosystem remnant influence how much of the remnant is not affected by a particular edge effect.
- Learning what the terms "mosaic" and "matrix" mean and how they may have an impact on the biodiversity in an area.

## *Major Points of the Chapter*

Habitat fragmentation affects different species in a variety of ways. For example, with increasing fragmentation (up to a point), edge increases. This can result in a decrease in edge-sensitive species and an increase in edge-generalist species. Likewise, the land use surrounding protected areas (the matrix) often can determine what biodiversity exists within the protected areas. Students need to become conversant with these ideas to ensure better management practices when they become professionals.

## Items That May Come Up While Teaching This Chapter

Students may become aware that virtually everything humans do fragments landscapes. As fragmentation becomes pervasive across a landscape, the natural heritage of that area will change. We hope that students will begin to realize that they need to think at the landscape level when managing natural resources. By so doing, they may begin to accept that they cannot restrict their actions simply to protected areas or public lands where they may work. They will have to find ways—different ways—to interact with diverse stakeholders and complex administrative boundaries if they are to be truly successful managing for natural resources. This approach is quite different from traditional management approaches of the past.

## Supplementary Activities

Design a car route that students drive on their own or in small groups. Along the way, they can itemize land uses they observe and how these uses fragment landscapes. Which ones seem to be most pervasive? Do some have greater impacts than others (roads versus rural homes)? What is the matrix where they live? What wildlife do they see; are they all edge-generalist species, or do they see any species that occur only in large, unfragmented landscapes? This activity will let students realize that these topics are not academic but truly relevant to where they live.

## Conducting the Exercises

### EXERCISE 8.1

#### How to Introduce or Conduct the Exercise

This is a thought exercise. You could ask students to write answers to these three questions, turn them in, and summarize the results to give back to the students. Or, they could generate their responses in class for collation on a chalkboard.

#### What to Expect When Students Do the Exercise

This exercise will make students think. By reporting their answers back to the class, you can facilitate learning and understanding by the class. This is so, we suspect, because it will reveal how little students actually understand about where they live, including how it looked before humans arrived. Some students will be confused by what these terms actually mean, now that they are asked to list examples. Their confusion can be turned around to help them understand terms such as "matrix" and how different land uses result in habitat loss, fragmentation, and degradation of landscape quality.

### Desired Outcomes for the Exercise

Students should connect land use and how it results in habitat loss, habitat fragmentation, and degradation of landscapes (e.g., increased soil erosion, altered hydrology, increase in invasive species).

### Points to Be Made Based on the Exercise

- Land use results in habitat loss, habitat fragmentation, and habitat degradation.
- Some land uses that result in these three factors are more pervasive than others.
- A single land use may result in all three types of losses.

## EXERCISE 8.2

### How to Introduce or Conduct the Exercise

Form student groups and have them visit the Species in Parks: Flora and Fauna Databases Web site (http://ice.ucdavis.edu/nps/) and compose species lists from national parks of different size. Have them plot the number of species versus park size and turn in their figures with answers to the questions asked.

### What to Expect When Students Do the Exercise

The figures should show an increasing number of species as the sizes of the parks increase, up to a point. The results will depend on what taxa they chose and the range of park sizes.

### Desired Solution for the Exercise

An appreciation that as protected areas get larger, species richness increases, and that the shape of the curve varies with taxa and the range of park sizes used.

### Points to Be Made Based on the Exercise

- Bigger is better when it comes to protected areas.
- Not all taxa respond the same way to size of protected areas. An area that supports all species of plants may be quite smaller than what it would take to protect all species of mammals.

## EXERCISE 8.3

### How to Introduce or Conduct the Exercise

Have the students address this exercise working in groups for the scenario they are using. They should list land uses that abut protected areas in the scenario and rank them by level of threat. They should also list species that will be most affected by these land uses. Finally, they should suggest one or two management actions that would minimize these threats. They can report their findings to the class.

### What to Expect When Students Do the Exercise

Students realize that answers to these questions are often not very straightforward. Likewise, they may learn that the land uses with the greatest threats to biodiversity are the very land uses that they participate in (homes, recreation).

### Desired Outcomes for the Exercise

An appreciation for the diverse array of land uses that abut protected areas and the depth of their effects on biodiversity.

### Points to Be Made Based on the Exercise

- Land uses adjacent to protected areas can affect what occurs within the protected area.
- These land uses are diverse and result in different types of edge effects.

## EXERCISE 8.4

### How to Introduce or Conduct the Exercise

Initiate a discussion in class based on the protected areas in the scenario they are working on and how such areas serve in protecting edge-sensitive species based on area (relative, because there is no scale), shape, and amount of edge.

### What to Expect When Students Do the Exercise

Students may realize that protected areas often were set aside with little consideration of how their area and shape may predispose them to not be ideal for edge-sensitive species.

### Desired Outcomes for the Exercise

An appreciation for the importance of shape and area when designing future protected areas so as to promote the viability of edge-sensitive species.

### Points to Be Made Based on the Exercise

- Shape and area will affect the amount of edge of a protected area, which, in turn, can affect the viability of edge-sensitive species.

## EXERCISE 8.5

### How to Introduce or Conduct the Exercise

Form students in groups and for the scenario they are working on, have them devise a list of management steps they would take to increase area, decrease the amount of edge, and decrease the isolation of areas that currently support biodiversity of interest. Each group also needs to prepare a timeline showing when management steps would be implemented. Groups can report their ideas to the class.

### What to Expect When Students Do the Exercise

Students may be stymied on how to proceed. That is to be expected, because they have little experience in management. Encourage them to think outside of the box. They may list ideas that are prohibitively expensive and are not realistic. Encourage them to think of these ideas anyway, but to then reexamine all their ideas once they have a long list. Do encourage them at that point to be pragmatic and realistic.

### Desired Outcomes for the Exercise

A list of management actions that realistically could be accomplished to make the scenario they are working on less fragmented and more connected than the way it is now.

### Points to Be Made Based on the Exercise

- Solutions do exist for reconnecting landscapes and decreasing fragmentation, but they usually are not easy. They require time, cooperation among diverse landowners and stakeholders, and resources.
- Conservation actions to decrease fragmentation should not be thought of just for public lands; they can also include privately owned land.

## EXERCISE 8.6

### How to Introduce or Conduct the Exercise

Break the class into small groups, and for the scenario they are working on, prepare a table listing the elements (human and natural) that make up the mosaic of the scenario. Students can share their lists with the rest of the class to make a complete list.

### What to Expect When Students Do the Exercise

Students will list a number of the elements that occur in the scenario; they will have uncertainty over a variety of elements, not knowing whether they are actual elements of a mosaic.

### Desired Outcomes for the Exercise

Students should realize that the mosaic of an area includes both human and natural elements. This is a more realistic and comprehensive appraisal than just being aware of the protected areas in public ownership. In addition, students may be surprised at how much of the scenario consists of human rather than natural elements.

### Points to Be Made Based on the Exercise

- A mosaic consists of all elements of a landscape—human and natural.

## EXERCISE 8.7

### How to Introduce or Conduct the Exercise

Students should think out the answers and then share them with the class.

### What to Expect When Students Do the Exercise

Students may not correctly guess that the matrix where they live is urban; they may try to look for a natural ecosystem. Likewise, the students will have difficulty trying to determine the matrix of the scenario they are working on. This is because the scenarios are composed of many different elements, making it difficult to actually determine the matrix.

### Desired Outcomes for the Exercise

Students should appreciate that humans have altered landscapes considerably and that actually being able to determine the matrix of an area can be difficult. They may also be surprised to learn that the matrix where they live is dominated by humans rather than by natural elements.

### Points to Be Made Based on the Exercise

- It is necessary to know what the matrix of an area is for effective natural resource management.
- Knowing the mosaic of an area facilitates knowing what the matrix is, particularly for landscapes with complicated administrative boundaries.

## EXERCISE 8.8

### How to Introduce or Conduct the Exercise

Form the class into groups, and have each group meet and address the challenge posed by the exercise. This exercise will require more than a few minutes, so you may wish to have students meet outside of class. After they have assembled their recommendations, it would be ideal if there was time in class for each group to present its recommendations.

### What to Expect When Students Do the Exercise

Groups will have to spend time discussing this challenging exercise, because the solution will have to be based on more than just science. We suspect their responses will vary widely; the more creative groups will come up with broad-based and long-term management actions that may include everything from land-use planning suggestions to the purchase of conservation easements and the formation of a local land trust.

### Desired Outcomes for the Exercise

The ideal product would include a comprehensive approach to the issue, would look across the diverse administrative boundaries of the scenario, would involve diverse stakeholders, and would have a timeline that covered several decades. An inappropriate solution would be one that simply said "buy the land and set it aside" or suggested that "county government should institute strict land-use planning that prohibited any further development." These suggestions are far too simplistic for the complex reality of the exercise. Any plans that showed an appreciation by the students of the difficulty of this exercise, and that attempted to deal with these challenges, would be acceptable.

### Points to Be Made Based on the Exercise

- The natural tendency of being human is to fragment landscapes; it is much easier to fragment than to reconnect an area.
- Success in slowing down fragmentation, or even reversing it, requires working cooperatively with diverse landowners and stakeholders; there is no simple way.
- One must think across administrative boundaries and think over long periods of time when dealing with fragmentation issues at the landscape scale.

## *Teaching Devices to Help Illustrate the Chapter*

- Show students figures that represent fragmentation over time, and demonstrate to them that the amount of edge increases only up to a point. When half or more of the landscape has been fragmented, the amount of edge will actually start to decrease.
- To illustrate that matrix can include both human land uses and natural ecosystems, ask students to list the matrix of cities, national parks, refuges and forests, and the lands between (private lands, usually in some form of agriculture).
- Have students list the names of common wildlife species, ones that most people know. Then ask them to categorize these species as being area-sensitive, edge-sensitive, dispersal-sensitive, or edge-generalist. Doing this brings home the point that most species we are familiar with (not the large charismatic species we see in zoos or in *National Geographic*) are edge-generalist species. Then ask them to list species they have seen in the wild that would qualify as area-sensitive, edge-sensitive, and dispersal-sensitive.

## Sample Test Questions

1. Three characteristics of habitat fragments are area, shape, and connectivity. With these in mind, define and discuss the species-area curve, area-sensitive species, edge-sensitive species, and dispersal-sensitive species.
2. How can the increasing isolation of a forest remnant reduce its biodiversity? Discuss how the shape of the forest remnant can influence the biodiversity of that remnant. Discuss the microclimatic changes often associated with the formation of an edge.
3. Matrix and mosaic are sometimes used interchangeably. Distinguish the two terms. What do each measure? Do both acknowledge human elements of a landscape? Which, in your opinion, is most useful in guiding conservation work?

## Supplementary Web Sites

- http://www.consecol.org *Conservation Ecology*, an online journal, publishes articles on fragmentation.
- http://www.conserveonline.org This site, produced by The Nature Conservancy, is a collection of practical information for students interested in conservation, including journal articles, unpublished reports, case studies, presentations, maps, and software. The documents can be downloaded or viewed online. Students can rank resources and write reviews.
- http://www.fs.fed.us/library/ This U.S. Forest Service library of publications and databases is available to anyone. Although you cannot search by keyword, the database is categorized by type of publication or data set. There are data sets pertaining to, for example, fire management, climate change, and regional flora—as well as maps and Forest Service software for downloading.

CHAPTER 9

# Managing Biodiversity Across Landscapes: A Manager's Dilemma

## *Learning Objectives for the Chapter*

THIS CHAPTER PROVIDES STUDENTS WITH A VARIETY OF APPROACHES TO MANAGING biodiversity across landscapes in diverse administrative ownership, from ecosystems where humans live to protected areas where humans only visit. Specific learning objectives include:

- Understanding that protecting biodiversity on nature reserves and protected areas is preferable to intensely managed enclaves, such as zoos, botanical gardens, butterfly pavilions, and aquariums.
- Understanding that because it will be next to impossible to protect enough land to ensure the maintenance of biodiversity and natural ecological processes, we must find ways to promote healthy landscapes where people live and work, the private lands of any nation.
- Understanding that ecosystem management works to ensure that entire landscapes, with their mixture of administrative boundaries, are more friendly to biodiversity and ecological processes. This approach seeks to blend the needs of biodiversity and humans and, ultimately, integrate the two rather than keep them apart.

## *Major Points of the Chapter*

Successful natural resource managers acknowledge that landscapes that harbor biodiversity, that allow ecological processes to occur, and that provide amenities and commodities for humans are necessarily complex because of diverse administrative boundaries. These managers are comfortable with the realization that there is not one way of doing business but a multitude of different approaches and styles for successful land management. Students need to be aware of this diversity, in order to be productive and successful in their chosen career. To do so, however, they need to become aware of these complexities. This chapter explores a variety of ways that students can span spatial and temporal scales to become contemporary natural resource managers.

## *Items That May Come Up While Teaching This Chapter*

Many students in natural resource management may have had an education that emphasized public land management. This is inevitable, because natural resource disciplines have focused on amenity and commodity uses that occur on public lands and protected areas. In addition, many of the present and historical controversies dealing with natural resources have focused on public lands. Conversely, much of many nations' land ownership is tied up in private ownership, necessitating an acknowledgment that public lands are not sufficient to protect biodiversity and ensure that ecological processes persist. This chapter (and others in this book) will enable instructors to partially address this imbalance in students' education. To do so, however, the teacher will have to repeatedly emphasize that ecosystem management acknowledges complex administrative boundaries and land ownerships, but also that different ways of natural resource management have to be utilized when working on lands not in public ownership. Students will be amenable to these complexities because they have not yet embarked on their careers.

## *Supplementary Activities*

A variety of teaching activities can increase student awareness of complex land ownership in implementing ecosystem management plans. For example, have students attend meetings of county commissioners and county planning committees. Arrange presentations by nongovernmental and governmental organization members who deal with natural resource management on private lands or across administrative borders. The Natural Resources Conservation Service and The Nature Conservancy are just two of these types of organizations.

## *Conducting the Exercises*

### EXERCISE 9.1

#### How to Introduce or Conduct the Exercise

This is a brief class exercise where you can ask for students' input while you list their answers on the board.

### What to Expect When Students Do the Exercise

Students will appreciate that species of interest in the scenario occur on lands with diverse ownership patterns.

### Desired Outcomes for the Exercise

Strive for a list that makes it obvious that students will not be able to ensure the maintenance of species of interest by simply working with the protected area managers.

### Points to Be Made Based on the Exercise

- Protected areas do not represent the full occurrence of species of interest.
- Ensuring the continuance of species of interest will necessitate working with private landowners and organizations other than the "traditional" groups concerned with protecting biodiversity.

## EXERCISE 9.2

### How to Introduce or Conduct the Exercise

This exercise can best be accomplished by forming groups of students and asking them to devise management strategies that employ both coarse-filter and fine-filter approaches for managing biodiversity. Each group could address both approaches, or you could assign some to one and the rest to the other approach. After convening, the groups could present their findings in front of the class and a synthesis could be developed that highlights the diversity encompassed by the two approaches, as well as strengths and weaknesses.

### What to Expect When Students Do the Exercise

Confusion will arise over the boundaries of each approach when the groups are in discussion. By working in groups on a specific landscape, students will come away with a more realistic understanding of what the two approaches address.

### Desired Outcomes for the Exercise

Students will not fully understand the point of this exercise until the groups report to the class and until following your attempt to summarize their findings. One expectation would be for them to appreciate that, in reality, both approaches need to be used to fully protect biodiversity. In addition, this would be an appropriate time to emphasize the importance for them to think across administrative boundaries, think outside of the box, and appreciate the temporal and spatial scales on which management actions lay.

### Points to Be Made Based on the Exercise

- Both approaches are useful and should be integrated in a comprehensive management approach.

- Spatial and temporal scales are important when designing management plans.
- One needs to think outside of the box, be creative, and attempt to include other groups and individuals to make either approach work.

## EXERCISE 9.3

### How to Introduce or Conduct the Exercise

Form the class into small groups. For the scenario they are working on, have each group develop landscape-level actions that could be undertaken to enhance the capacity of the existing protected areas to maintain biodiversity. Encourage students to think out of the box but to be specific rather than vague in their suggestions.

### What to Expect When Students Do the Exercise

The students should have little trouble coming up with ideas. They may, however, have difficulty with their ideas passing the test of "practicality." The most common suggestion will be to buy private land adjacent to protected areas. Whether or not this is practical will depend on the assumptions they give you.

### Desired Outcomes for the Exercise

The point to emphasize in this exercise is that the existing protected areas in the scenario are probably not adequate to ensure the maintenance of biodiversity. To make them better (bigger, better shaped, more connected and less isolated, etc.), they will have to think across administrative boundaries to find workable solutions. This approach will require them to think originally and test a variety of ideas, ranging from land-use planning at county and state levels, to legislation at state and federal levels that may provide economic incentives for private landowners to keep their land out of development or enhance it for biodiversity, to working with land trusts or government organizations. Encourage them to stretch their minds and consider a variety of approaches that may have promise.

### Points to Be Made Based on the Exercise

- Protected areas are often not the right size or shape, and they may not have the appropriate level of connectivity to other protected areas to adequately maintain biodiversity.
- Altering this situation requires more imagination than simply saying additional areas should be purchased and added to the existing protected areas. Innovative approaches may include legislation, economic incentives, and other unlikely approaches that natural resource managers have not traditionally employed.

## EXERCISE 9.4

### How to Introduce or Conduct the Exercise

Ask the students for answers to these questions, listing them on the board. This is best done for the scenario they are working on. Perhaps you could ask each student to provide a one-paragraph summary of the ideas presented, to be turned in later.

### What to Expect When Students Do the Exercise

By answering these questions, students will come away with a better understanding of the scenario they are working on. They should realize that the existing protected areas are not connected by movement corridors, and if they are ever going to be connected, it will require working with private landowners.

### Desired Outcomes for the Exercise

A heightened awareness of how little of the landscape is actually protected and that the existing protected areas are unduly influenced by the surrounding private lands and their uses.

### Points to Be Made Based on the Exercise

- Landscapes are fragmented, and protected areas are seldom adequately connected with movement corridors.
- It will take creative thinking and hard work to reconnect increasingly fragmented landscapes.

## EXERCISE 9.5

### How to Introduce or Conduct the Exercise

Form the class into groups, and have them design a movement corridor connecting two protected areas in their scenario. Ask them to prepare suggestions for each of the questions in the exercise. Groups can report back to the class or turn in written responses.

### What to Expect When Students Do the Exercise

You will receive a variety of group answers, ranging from ones that are quite incomplete to (perhaps) ones that would be a credit to the most creative natural resource manager. By the level of detail in their answers, it will be evident whether the group truly collaborated on their response. Whether they adequately incorporate adaptive management into their response will also serve as evidence of how well they understand this concept.

### Desired Outcomes for the Exercise

This exercise should encourage students to begin to think as future natural resource managers and practicing conservationists. When confronted with a real landscape that is inherently fragmented by diverse ownership, they will have to be flexible in their thinking and creative in their imaginings to stitch back together landscapes that have been fragmented.

### Points to Be Made Based on the Exercise

- Designing movement corridors to connect protected areas requires creative thinking. Because these intervening landscapes are often in land uses other than protection and owned by individuals rather than government agencies, it will be extremely challenging to develop workable plans for movement corridors.
- Adaptive management is essential when designing movement corridors. If it is not employed, there will be little learned, ranging from the short-term knowledge of whether the corridor is being used at all, to the longer-term benefits regarding how future movement corridors can be better designed.

## EXERCISE 9.6

### How to Introduce or Conduct the Exercise

Form the class into groups, and have them prepare a report that addresses the questions in the exercise for their scenario. In addition to the table they will prepare that lists the ecosystems and land uses of the protected areas and adjacent lands, ensure they answer the questions posed at the end of the exercise.

### What to Expect When Students Do the Exercise

Some of the information asked in the exercise may not be available in the scenario description; however, students should not be deterred from providing the information that is available. They may have to provide certain pieces of information based on their best understanding of the scenario description. After preparing the table, the students should be challenged to think creatively when answering the questions the exercise asks them.

### Desired Outcomes for the Exercise

Strong contrasts should be evident when comparing land uses adjacent to the protected areas. This exercise will challenge the students to acknowledge the magnitude of the task when working across administrative boundaries to protect biodiversity and ecological processes. At the same time, it may stimulate them to think of creative ideas that might dim the stark contrasts in land uses between protected areas and land in private ownership. These contrasts will not be as severe for all land uses on private lands. Where the differences are not so great, the opportunities for collaboration are enhanced.

### Points to Be Made Based on the Exercise

- A landscape-level assessment of biodiversity protection must address land uses and ownership adjacent to protected areas.
- An effective conservation approach will be aware of these differences in land use and ownership and attempt to find ways that minimize adverse affects.
- These types of efforts, when done well and given enough time, may promote conservation on private lands and minimize contrasts between protected areas and adjacent areas.

## EXERCISE 9.7

### How to Introduce or Conduct the Exercise

Lead a class discussion for a species in your scenario that addresses these questions. Afterward, have each student turn in a one-paragraph description summarizing the strengths and weaknesses of Habital Conservation Plans (HCPs).

### What to Expect When Students Do the Exercise

There will probably be a great deal of confusion and uncertainty in their responses. This would be a great opportunity to send students to the Web sites on HCPs and related issues to clarify aspects of this federal private lands initiative to protect threatened and endangered species within the Endangered Species Act.

### Desired Outcomes for the Exercise

This exercise will work if students come away with an appreciation of efforts by the federal government to find ways to protect threatened and endangered species on private lands.

### Points to Be Made Based on the Exercise

- HCPs are the latest attempt to find ways that allow the Endangered Species Act to work on private land.
- There are limitations on how well HCPs can work on private lands, because HCPs assume both adequate resources and sufficient knowledge by the private entities that are responsible for implementing the HCPs.
- Adaptive management is a critical step in modifying and learning from HCPs.

## *Teaching Devices to Help Illustrate the Chapter*

- Have students visit Web sites to learn more about the complexities, successes, and failures of HCPs and related approaches to conservation on private lands.
- Table 9.1 lists steps to protect and manage for biodiversity under the

rubric of ecosystem management. Either in class or as an out-of-class assignment, have the students critique this list. Encourage them to make it better, eliminating weak points and adding missing items.

- Have students pick an area that has at least some protected areas, and then go online to create maps showing the complexity of political boundaries. The point of the exercise would be to illustrate the reality, and the complexity, of managing for biodiversity and ecological processes at the landscape level. *Online Map Creation* (OMC) is an amazing resource for anyone needing to make a map. Students can create Postscript-formatted maps of just about any locality by inputting geographic coordinates. OMC is a subset of GMT (Generic Mapping Tools) software developed by scientists at the University of Hawaii and NOAA. A variety of projections can be made, including political boundaries, rivers, and topography. Step-by-step instructions are provided.

## Sample Test Questions

1. Contrast the coarse-filter and fine-filter approaches for protecting biodiversity by listing the strengths and weaknesses of both. Make a case for why a comprehensive approach to protecting biodiversity should use both.
2. Any particular landscape has a mixture of protected areas, as well as areas in human land uses and ownerships that are not compatible with certain ecological processes or biodiversity. List a variety of strategies a conservation practitioner could pursue to minimize the contrasts between protected areas and the other lands that would tend to make the overall area more conducive to ecological processes and biodiversity.
3. List and discuss the steps you would follow to design and evaluate a movement corridor for a species that required corridors to disperse from one protected area to another. What general recommendations can you make regarding gap width in corridors in relation to speed of animal movement, corridor width and surrounding land use, and corridor width and length?
4. Contrast natural resource management that restricts its activities to within the boundaries of protected areas and an approach that works across administrative boundaries. Which would be easier? Which might be more effective? Which approach requires more diverse and complicated approaches?
5. Describe how Habitat Conservation Plans (HCPs) work on private lands. How does this approach differ from managing endangered and threatened species on public lands? Describe one strength and one weakness of HCPs.

## Supplementary Web Sites

- http://endangered.fws.gov/hcp/index.html. The U.S. Fish and Wildlife Service provides incidental take permits for endangered species and their habitat to landowners who mitigate for adverse effects through Habitat Conservation Plans. This Web site provides up-to-date changes in HCPs, as well as success stories and examples.

- http://www.endangered.fws.gov/landowner/index.html. This Web site of the U.S. Fish and Wildlife Service discusses safe harbor agreements, which offer assurances to landowners who enhance their property that attracts endangered and threatened species.
- http://www.environmentaldefense.org/safeharbor This Web site of Environmental Defense discusses safe harbor agreements and provides comments and related resources on this issue.

CHAPTER 10

# Working in Human Communities

## *Learning Objectives for the Chapter*

THIS CHAPTER INTRODUCES THE CONCEPT AND SKILLS OF WORKING WITH PEOPLE TO accomplish our conservation mission. Specific learning objectives include:

- Understanding the necessity of working with people as part of natural resource conservation, especially in an ecosystem context.
- Developing an understanding of and appreciation for listening skills.
- Developing skills in characterizing and satisfying stakeholders needs.
- Understanding that collaboration is the only practical way to make progress in ecosystem management.
- Encouraging an appreciation of the many ways in which stakeholder involvement can occur.
- Appreciating the fact that experts are needed to accomplish successful stakeholder involvement.

## *Major Points of the Chapter*

The major point of the chapter is that ecosystem management must occur with the participation of the people who live in the ecosystem—it is not something created by experts and imposed by agencies.

Direct coverage of stakeholders is contained in only one chapter, whereas the biological aspects and process aspects cover several chapters each. This does not mean, however, that working with stakeholders is less important than the other topics. In fact, one of the defining characteristics of ecosystem management is that it seeks to work with and through communities rather than apart from them. We suggest, therefore, that this chapter be given ample time and attention during the course.

## *Items That May Come Up While Teaching This Chapter*

Students rarely choose to major in natural resources because they like people; rather, they like nature and may prefer to not deal with people. Thus, some students may regard this chapter negatively; it may seem unnecessary or an interference with the things that are important to them—the biological things. Teaching this chapter (and throughout the course), instructors will need to emphasize that working with people is important, reinforcing the concept with their own experience and stories. This is, we feel, perhaps the most important chapter in the book.

Students may also object to the idea that different stakeholders need to be treated differently. They may take the view that we are all people who should enjoy the same access to information, to decision makers, and to decision making. It is important to emphasize that the chapter asks them to think (somewhat) realistically about what happens when working with people, rather than (completely) idealistically. This is often a difficult concept to work with in a university setting, where we are all considered equal.

Instructors may feel somewhat out of their depth in this chapter, as it deals with human sociology and psychology. This, then, is a perfect place to suggest to students that we are all learning together, that when discussing humans and their behavior, we have to be fuzzier. There are few "rights and wrongs" in this chapter.

## *Supplementary Activities*

A field trip to a public meeting or other form of stakeholder involvement makes a perfect supplement to this chapter and may prove to be the most important educational experience students can have in ecosystem management. A local planning board meeting or city council meeting can be just as effective as a state fish and wildlife agency meeting, so don't overlook what may be happening locally. It is often useful to have a debriefing session with an official from the meeting, to talk about what happened and why.

Mock public meetings can also be useful, involving the students in role-playing in their scenario. Students who wish to pursue special studies might be invited to develop and staff a mock stakeholder involvement event, using one of the techniques in the chapter and one of the issues from the scenario.

A guest speaker who specializes in conducting stakeholder analyses for a state agency is a great addition here. He or she can speak effectively about what works and does not work well in specific situations. A legislator or legislative aide can be very effective talking about the realities of who gets listened to and who does not.

Box 10.3 discusses the trust responsibilities of federal agencies. This is a difficult topic to nail down, because it is fuzzy even in the federal statutes. If a federal representative and/or tribal representative are available, he or she would make great guests.

Box 10.4 describes nine attitudes toward nature, developed by Steve Kellert of Yale. Creating a profile of student attitudes in class can provide a useful way to show students how broad the range of attitudes is. Ask students to list the three attitudes that they hold most strongly, and compile their lists to show the class profile. Then compare these with approximate figures for the general U.S. profile: aesthetic—15%, dominionistic—5%, humanistic—35%, moralistic—20%, naturalistic—10%, negativistic—25%, scientific—1%, symbolic—10%, utilitarian—20%. (They add to more than 100, of course, because people hold different views.)

Box 10.5 describes the nominal group technique, the most common and easiest form of group process in use today. Conducting a nominal group process is a very interesting activity for a student who wishes to do extra work. It can be done for a topic within the department (e.g., the most important addition to our educational program would be . . .) or the course (e.g., I would like to have learned more about . . .). Alternatively, students could conduct a nominal group process for their scenario, using trigger questions that relate to the scenario as a whole (e.g., for the ROLE Model to succeed, we most need to . . .) or for a particular topic (e.g., to do a useful assessment of the possible results of relocating the Camp Fraser waste dump, we would need to know . . .). Conducting such an exercise will show students how valuable it is to have a structured process that will get to a desired product.

Universities usually have a program that trains facilitators and then offers them for use by university groups that need facilitation help. Bringing a trained facilitator in to conduct this exercise would also show how effective it can be, reinforcing the idea that expert help is usually needed.

Box 10.8 discusses the art of listening, as does the text. Universities also usually have experts on listening skills in the counseling or study skills program. Bringing an expert on listening skills to class would be useful, especially if the person brings some exercises regarding listening skills.

## *Conducting the Exercises*

### EXERCISE 10.1

#### How to Introduce or Conduct the Exercise

Designed to be done in small groups, this exercise should take only a few minutes.

#### What to Expect When Students Do the Exercise

Students will probably have great empathy for the farmer and none for the golf course/condominium owner. They will probably start talking about “they”—

they should help them, they should pay for it. Ask students who "they" are, probing to get them to understand that "they" are us.

### Desired Outcomes for the Exercise

The exercise is designed to help students see that what should be done in any situation often has subjective elements to it—that we sympathize with some people and not with others. There is no right answer; the desired outcome is for students to get into the heart of the matter by seeing the complexities.

### Points to Be Made Based on the Exercise

- Who the people are changes the way we feel.
- Our own value system plays an important role in how we think.

## EXERCISE 10.2

### How to Introduce or Conduct the Exercise

See the general description for using the scenario exercises. This is a multiple-part exercise that asks students to (1) identify many stakeholders of different levels of intensity, (2) analyze a few of the more intensely interested stakeholders, and (3) consider broader questions about these intense stakeholders. Close reference to the text will be needed to keep the ideas of stakeholder orbits and stakeholder assessments in mind. Depending on the time available and the size of the class, parts 2 and 3 might need to be restricted to just one stakeholder group.

### What to Expect When Students Do the Exercise

This exercise usually gets students very involved, and they enjoy the work. Students get to make judgments about people's interests.

Students may not wish to make judgments that some people are more important than others. Encourage them to select low-orbit stakeholders on the basis of which stakeholders might self-select to be intensely involved, emphasizing that the stakeholders choose their level, not us.

Students may also just answer with the first ten stakeholders that come to mind. Encourage them to keep thinking to get a good array of stakeholders at all orbits.

### Desired Outcomes for the Exercise

The desired outcome is for students to get to know some of the stakeholders in their scenario. The practice that they get thinking about the stakeholders in depth will help them begin to understand that people are different and need to be worked with differently.

### Points to Be Made Based on the Exercise

- Stakeholders differ in their interests and the intensity of their interests.
- Careful analysis of stakeholders will reveal clues about working with them effectively.
- We might well disagree on how important various stakeholders are; that's why stakeholders make their own decisions about how to participate

## EXERCISE 10.3

### How to Introduce or Conduct the Exercise

See the general information on using the scenario exercises. Also, tell students that this is an open-ended exercise in which they will have to use much judgment and knowledge about both stakeholders and the issues in the scenario. Refer them to the levels of involvement in the chapter, as this is the structure for the exercise.

### What to Expect When Students Do the Exercise

This is the major exercise for this chapter. It will require considerable time, probably distributed over several class sessions or days, so students can work their way through it. Because it is so open-ended, student responses will be very different, and some groups may need considerable help working through the exercise.

### Desired Outcomes for the Exercise

The desired outcome is really for the student groups to design a small stakeholder involvement program around the assigned issue. To that end, students should identify actions, stakeholders, and techniques (or at least approaches) for working with the stakeholders. As with all exercises in this chapter, however, there is no right answer. The important outcome is for students to have seriously addressed the stakeholder involvement process.

### Points to Be Made Based on the Exercise

- Different kinds of actions require different kinds of stakeholder involvement.
- Working with stakeholders is hard—it takes lots of time and judgment.
- The outcome is uncertain; although planned to occur one way, stakeholder assessment may turn out another way.

## EXERCISE 10.4

### How to Introduce or Conduct the Exercise

See the general description on using the scenario exercises. This should be a relatively quick exercise, which can be expanded or contracted to fit the available time.

### What to Expect When Students Do the Exercise

Students may have some difficulty with the need to combine stakeholders into homogeneous groups for several reasons. First, they will need to fight the normal tendency to want to make groups that have all opinions present. But this is the nature of focus groups; emphasize that homogeneous groups allow opinions to be fully expressed, rather than repressed.

Second, students may have trouble defining what are the homogeneous characteristics to use in grouping stakeholders into focus groups. Should they group by kind of resource (fish, game, endangered species); resource use (recreation, hunting/fishing, mining, grazing); resource value (money, food, tradition)? This is a major learning element. Encourage students to group in any way that makes sense for the question at hand, and remind them that not all groups need to be parallel. One could be businesspeople, another could be recreation interests, whether providers or users of recreational services.

Third, students may run out of stakeholders to make up certain focus groups. Encourage them to expand on the list of stakeholders they know by adding those that they expect would also want to be included in the focus group.

### Desired Outcomes for the Exercise

Students should be able to group stakeholders into reasonable sets and to generate statements that would help the members express themselves. As with all exercises in this chapter, there is no right answer; very different groupings will be just as useful to the process of learning about stakeholder interests and values.

### Points to Be Made Based on the Exercise

- The focus group is a specific type of stakeholder involvement that works well under specific conditions.
- Grouping stakeholders into subsets is a subjective task, with no right answer.
- Groupings made for one issue may not be useful for another issue.

## *Teaching Devices to Help Illustrate the Chapter*

We have often used a simple puzzle to get students to understand that there are many ways to look at a question, generating lots of different answers, and that working together is a good way to get to agreement. We begin this material by asking students how many squares are in the following figure:

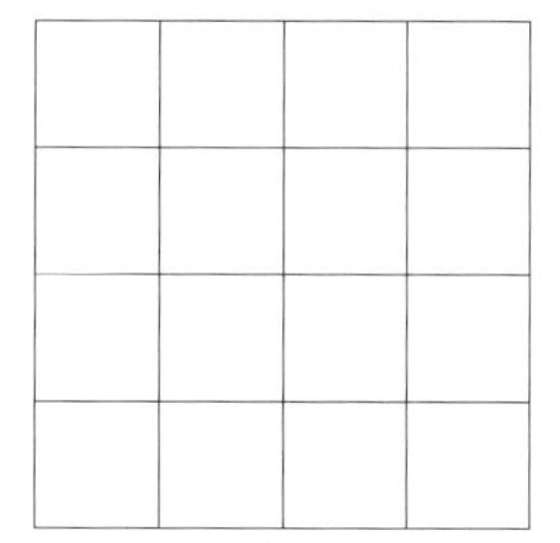

They shout out many answers, including 1, 16, and 17, and then a variety of answers in the 20s. We write each one down and ask how there can be so many answers to such an easy question. We then ask people to vote for which one is right and tally the answer. We then ask if taking the number with the higher votes means that it is the right answer.

Finally, we ask students to explain how they got to their answer. As the dialog continues, we eventually get to the point that this is a matrix, with squares that are $1 \times 1$, $2 \times 2$, $3 \times 3$, and $4 \times 4$; there are 16, 9, 4, and 1 of these, respectively, for a total of 30 squares. At this point most students are recognizing that they have talked themselves to a new, better understanding that yields a pretty good answer.

Then we ask them how many squares there would be in a $5 \times 5$ matrix, and the answer of 55 emerges rapidly (recognizing that this is the sum of squares, they add 25 to the original 30). Then, we go on to a $6 \times 6$ matrix, for which the answer is 91.

The purposes of this exercise are to get students to see that they do not necessarily have the right answer, that conversation can help them view things differently, and that after they have worked together on this easy problem, the answers to subsequent more complex problems can come more easily. It is a perfect introduction to the value of working with stakeholders.

## *Sample Test Questions*

1. For a particular circumstance (supply the circumstance), describe how you would go about ensuring that all three parts of the success triangle were achieved.
2. Describe the five stages of stakeholder involvement and the circumstances under which each of them would be effective and/or appropriate.
3. Contrast the purpose, style, and participants in a nominal group technique and a focus group.

## *Supplementary Web Sites*

- http://www.cbcrc.org The Community-Based Collaboratives Research Consortium publishes materials about stakeholder involvement in natural resource issues. The Web site is maintained through the University of Virginia.

- http://www.mind.net/app The Applegate Partnership describes their efforts at community-based conservation.
- http://www.epa.gov/stakeholders The U.S. Environmental Protection Agency's site describes the EPA's public involvement program, including basic information about public involvement techniques and practices.
- http://www.fhwa.dot.gov/reports/pittd The Federal Highway Administration's site includes a pubic involvement manual.

# Strategic Approaches to Ecosystem Management

## *Learning Objectives for the Chapter*

THIS CHAPTER INTRODUCES THE CONCEPT OF STRATEGIC MANAGEMENT, THE PROCESS of thinking and acting explicitly about what a person or group is doing. Specific learning objectives include:

- Understanding the formal process of strategic management, in the form typically used by natural resource agencies and professionals.
- Appreciating the need to go beyond strategic planning to a more complete system of strategic management.
- Demonstrating how strategic management can work in a variety of circumstances, including the idea that this does not have to be as formal or formidable as it sounds.

This is also the major chapter in which students are asked to use the materials in earlier chapters to develop true ecosystem management plans and activities. Consequently, although the chapter is relatively short, it should be given extensive time in the course, so that students have an opportunity to integrate and apply the earlier material. Therefore, an additional and very important learning objective is:

- Applying the concepts of ecosystem management to develop actual ecosystem management missions, strategies, goals, and objectives.

## Major Points of the Chapter

The major point of the chapter is that explicit decisions need to be the basis for action in any part of life, but particularly in ecosystem management. A fundamental premise of this book is that the technical aspects of ecosystem management must be considered within a strategic approach.

Strategic planning is often considered a waste of time by academics (universities change very slowly), so teachers may actually bristle at some of the content and perspectives in the chapter. Please note that this chapter is about using strategic planning as the basis for action—a key concept for the chapter—not as a hollow exercise to write a nice plan.

## Items That May Come Up While Teaching This Chapter

Most students in universities will have relatively little experience with strategic management because they have not worked in situations in which strategic plans were written. It is important to make this material real by spending some time discussing decisions and processes in which students might have been involved.

Students who have some experience in business or government will probably have negative feelings about strategic planning because they have seen the process done poorly or incompletely, and because they have been at the lowest level of dealing with a strategic plan created by higher-level employees or executives. They may wish to complain about a situation in which they participated.

A common complaint is that someone has been involved in a strategic planning process in which they really wanted to make big changes, and they were rebuffed. The chapter points out that strategic planning is not a route for revolution, because decision makers will not endorse such. Cautioning students not to expect the world to change massively is important. Remember what Winston Churchill said: "A 20-year-old who isn't a liberal has no heart; a 40-year-old who isn't a conservative has no brain."

Students may get hung up on the terminology, arguing that the words in the chapter are not the same as they learned them. The important point is that everyone at any time and place understands what the words mean—not that these definitions are the ones that have to be used everywhere and for all time.

## Supplementary Activities

Having students review strategic plans from the university or local agencies is a good way to get them involved. Compiling a set of such plans for agencies, departments, and NGOs in the area can be very useful to give to students so they can see some real plans. Most organizations now have all or parts of a strategic plan on their Web sites, so students can investigate them independently.

Box 11.1 describes various forms of strategic planning that have been popular. The current vogue (and the current federal law) involves the GPRA

(Government Performance and Review Act). Many details regarding the GPRA are available on the Internet at http://www.whitehouse.gov/omb/mgmt-gpra/. An excellent special activity or extra credit would be a review of GPRA in regard to the principles and details described in the chapter.

Boxes 11.7 and 11.9 take strategic management to the final steps of developing problems, tactics, and projects. We have not included exercises relating to these steps because it would be an empty activity to make up details about fake projects. However, students could proceed through the entire process for a local ecosystem, community, or university program or activity. For courses that use a strong role-playing approach, teams could be formed to actually create criteria and judge project proposals.

## *Conducting the Exercises*

### EXERCISE 11.1

#### How to Introduce or Conduct the Exercise

This exercise is designed to be done either individually or in small groups. Ask students to think about some activities they have been involved in, either as individuals or part of a group. Something as simple as how they plan their week can work; also, New Year's resolutions, savings from their summer job, exercise plans.

#### What to Expect When Students Do the Exercise

Students may need to be prodded to think actively about this. Using one student as an example may help; ask him or her directly about something simple, like planning his or her week.

#### Desired Outcomes for the Exercise

The exercise is designed to help students see that they are engaged in strategic thinking all the time, whether they recognize it or not, and that their reactions to it vary according to many surrounding situations (forced versus voluntary; formal versus informal; big change versus small change).

#### Points to Be Made Based on the Exercise

- We all are engaged in planning all the time.
- Most of the time, we do not follow through very well because it is hard to make our short-term desires be subsidiary to our long-term needs.
- In groups, planning requires lots of talk and compromise.

## EXERCISE 11.2

### How to Introduce or Conduct the Exercise

See the general description on using the scenario exercises.

### What to Expect When Students Do the Exercise

This exercise makes students confront the question of what data to collect and what to forget about. There are two parts: first, they need to set criteria for selecting data to be collected; second, they need to apply those criteria to the example for a scenario.

This will be difficult for them, because they probably have not been the decision makers with limited time or budget (it is easy to make a long list, but not easy to make the judgments). They may also get hung up on what specific data to collect (e.g., abundance of organisms) rather than on what criteria to use to judge what data to collect (e.g., information required by law). Providing some ideas might be needed, such as:

- Data that reflect the long-term condition, rather than the short term.
- Data that relate to extinction/extirpation.
- Data that represent items that have enormous impact on the system (e.g., winter snowpack).
- Data about how residents/leaders feel about things.

To make the situation more real, students could be given costs, in time or money, for various kinds of data, and a budget of how much time and money they have to spend.

### Desired Outcomes for the Exercise

The desired outcome is for students to make some tough decisions about what can and cannot be collected, as a first step toward other exercises in which they will have to be the decision makers.

### Points to Be Made Based on the Exercise

- This is a hard process, but is necessary because all resources are limited.
- Decisions can be made on a lot less information than we ordinarily think.
- This is just information, it is not the final product; the next exercises start getting to the real outcomes and benefits.

## EXERCISE 11.3

### How to Introduce or Conduct the Exercise

See the general information about using the scenarios.

### What to Expect When Students Do the Exercise

This will be an easy and enjoyable exercise for students. They will get to use big words and lofty ideals to write a mission statement. They may have some conflicts in their teams if some members want to get very specific and some want to be very general.

The usual pattern is for the first iteration to be quite short, then for people to start adding lots of phrases covering parts that they think are important, and finally for the statement to get whittled down to a short sentence or phrase again. However, different groups may come up with very short or very long statements—a good opportunity to point out that either one is okay, as long as it is meaningful.

### Desired Outcomes for the Exercise

Recognition that a mission statement is meant to be inspirational but does not get much further than that. Another desired outcome is a mission statement that the student group will use through the other exercises.

### Points to Be Made Based on the Exercise

- It is fun to write a mission statement, because it appeals to the reason for our getting into the field.
- Even writing a mission statement can bring disagreement, so it is necessary to compromise.
- Putting these words down, though, is relatively easy because most of us can see our interests covered in lofty phrases.
- Mission statements are meaningful in detail to those who wrote them, but they will be interpreted differently by everyone.

## EXERCISE 11.4

### How to Introduce or Conduct the Exercise

See the general description about using the scenario exercises. This is a short exercise and may be the scenario exercise that is most amenable to individual student work.

### What to Expect When Students Do the Exercise

This is a typical exercise for a university class. It asks students to give an example of five different approaches. It should go quickly and easily.

### Desired Outcomes for the Exercise

Students should be able to make clear distinctions among the five strategies listed. Also desirable is the understanding that several strategies may be used simultaneously.

### Points to Be Made Based on the Exercise

- Real differences exist among strategies.
- Selecting a strategy has important consequences on other later decisions. (Teachers may wish to speculate with students about what subsequent decisions they would make if using one strategy or another—would you build an education center if your strategy were technical assistance or incentives?)
- Strategies are often implicit rather than explicit, so you might need to infer them from a strategic plan or subsequent actions.

## EXERCISE 11.5

### How to Introduce or Conduct the Exercise

See the general description about using the scenario exercises. This is a two-part exercise. The first part is evaluating a given set of goals; this part is really more of a talk-about-it exercise. The second part is a true scenario exercise, asking students to write goals for a scenario.

### What to Expect When Students Do the Exercise

The first part of the exercise should go easily and perhaps should be handled as a general class discussion or short group discussion. The second part is one of the two major exercises in the chapter; the collaborative work will probably take considerable time—at least a full laboratory period. This will begin to challenge students, because they probably have not written goals before. Sometimes the process follows the same pattern as for writing a mission statement—first very general goals, then more specific goals, then a combination—expect that students will end up with a few goals that are still very broad.

Students may want more guidance on how much their goals are supposed to cover. Direct them back to their mission statement and ask them to make sure that several of the most important areas are covered. But also remind them that this is an exercise, and they do not have to be certain that everything is fully covered.

Students may want to make all their goals parallel (e.g., a goal about biota, one about habitat, and one about people; or, one about aquatic resources, one about terrestrial), rather than mixing them (e.g., adding to the earlier lists a goal about a diverse work force). But goals do not all need to be of the same scale or type; they should be chosen to highlight those things that are most important in broad thinking. If diversifying the work force is a compelling value, then it is perfectly appropriate, even if all the other goals are based on direct natural resource components.

Students may also be tempted to drop down to specific achievable items in the goal-setting stage. But goals are not meant to be achievable; they are meant to be aspirational and to have a long life. Keep students focused for now on what goals are in general; the next exercise will be more specific.

### Desired Outcomes for the Exercise

Students should prepare a list of goals that covers their landscape scenario thoroughly, especially those aspects that they or others think are important.

### Points to Be Made Based on the Exercise

- Goals begin to dissect the mission, providing more guidance.
- This begins to be hard work, because we start to see the form of the work that will eventually be done; things not covered now will not be turned into projects later.
- People start to strategize now to make sure their interests are covered.
- The goal list can still stay quite general, and another iteration (subgoals) may be needed to provide a bit more guidance.
- Different approaches to goals—a few big ones, or several smaller ones—are perfectly appropriate.

## EXERCISE 11.6

### How to Introduce or Conduct the Exercise

See the general description about using the scenario exercises. This is another two-part exercise. The first part asks students to evaluate some objectives; this is really a talk-about-it exercise, and it can be handled as a general or group discussion.

The second part is the other major exercise in the chapter, asking students to prepare objectives that fit under the goals they produced in Exercise 11.5. As it will require a considerable amount of time, it should be separated from the previous exercise by an interval.

### What to Expect When Students Do the Exercise

Students will find this exercise difficult because it is so hypothetical, but they just need to treat it as if it were as real as possible. Keep referring them back to their scenario—what they think are the most important items to work on. Also, keep them focused on the ecosystem management aspects of the scenario, rather than just doing regular resource management work (e.g., doubling harvest rates). Repeatedly ask them to think about the items in Chapters 5–10 and how they relate to the scenario.

### Desired Outcomes for the Exercise

Students should be able to write and evaluate objectives that are SMART. They should do an analysis of their objectives relative to each of the items that SMART stands for.

**Points to Be Made Based on the Exercise**

- This is where the rubber meets the road—where hard choices get made.
- When they set objectives, students are actually defining what they will and will not do. This will feel uncomfortable to them; assure them that this is what effective work is all about.
- Objectives do not need to be neat, one-sentence items; they can be sentences, lists, phrases, paragraphs—whatever is needed to convey the SMART message

## *Teaching Devices to Help Illustrate the Chapter*

Box 11.3 defines the terminology of strategic management that we recommend using for the course. Being precise about terminology is important, so that everyone understands one another. This point can be illustrated by asking students to write down the number from 0 to 100 that comes to mind when you say the following words: all (everyone will say 100); none (everyone will say 0); a majority (most people will say 51, but not everyone); most, many, a few, several, and some. After running through the same list, go back to the beginning and write on an overhead the students' answers as they shout them out. The reason for being precise will become very apparent!

In his book, *The Seven Habits of Highly Effective People*, Steven Covey presents drawings that depict an old woman and a young woman, and then a composite that includes the drawings of both at one time (pages 25, 26, 45). He uses the composite image to talk about how the information one has determines how that person views the world; we sometimes need a paradigm shift to get us to think differently. This device works well in class. Ask half the class to close their eyes and show one image to the other half of the class, then show the other image to the first half. Then, with all the eyes open, ask the class what they first see in the composite image. There will be mass confusion! Point out how to see the two images in the same picture. This is a useful device to begin the lesson, showing students that what they are preconditioned to see or believe influences what they actually do see. Talk about how strategic management is a different way to view the world, a perspective that will require them to shed some of their current ways of thinking.

The difference among the various levels in the strategic step-down is one of the most important lessons in the chapter, especially going from goals, which are general, to objectives, which are specific. The title song from the musical Camelot can be used to illustrate this idea. When King Arthur is describing Camelot, he is really describing a "Strategic Plan for Climate." The words of the song include several goals or subgoals (e.g., "the climate must be perfect all the year"), several objectives (e.g., "the rain must never fall 'til after sunset"), and even a mission statement for Camelot as a whole (i.e., " . . . a more congenial spot for happy ever aftering . . . "). Try playing the song for the students to hear while progressively revealing the lines of goals and objectives. Then, return to the list to show the distinction between the two levels. Wearing a Burger King crown during the music also helps!

Virtually all state fisheries and wildlife agencies have professional planners on their staff. Inviting the agency's planner to talk to students might give them a real-life glimpse into the promise and frustrations of these tasks.

## Sample Test Questions

1. Describe the purpose of strategic management and the characteristics that make it useful in ecosystem management.
2. Compare the concepts and characteristics of goals and objectives.
3. Characterize and then improve how the following objectives (give a series of objectives) are written, in terms of being SMART.

## Supplementary Web Sites

- http://www.chesapeakebay.net The Chesapeake Bay Web site describes this large and complex project.
- http://www.calfed.ca.gov The CALFED site similarly addresses another huge effort in planning.
- http://www.sfrestore.org The South Florida Ecosystem project, another large-scale ecosystem project.
- http://www.owpweb.org The Organization of Wildlife Planners is the professional group composed of leading planners from state and federal agencies in the United States. This group is an excellent source for information and perhaps resource people who can come to class.
- http://www.training.fws.gov/iafwa/mat.html The U.S. Fish and Wildlife Service operates a program called the Management Assistance Team, which helps state and federal agencies advance their management approaches.

# Evaluation

## *Learning Objectives for the Chapter*

THIS CHAPTER EXPANDS ON THE CONCEPT OF EVALUATION, WHICH WAS INTRODUCED IN Chapter 11 as part of strategic management. Specific learning objectives include:

- Understanding the purpose of evaluation as an aid to decision making.
- Contrasting the three styles of evaluation—formative, process, and summative—so students will be able to distinguish their uses, characteristics, and values.
- Appreciating the fact that evaluation plans need to be built into a project or activity from the very beginning, not tacked on at the end.

## *Major Points of the Chapter*

The major point of the chapter is that evaluation should be a formal and explicit part of action. Also, the chapter presents evaluation as another way to think about learning while doing; thus, it is another manifestation of adaptive management.

The chapter takes an academic approach to evaluation, treating it in the three forms that are usually described in an evaluation text or course. This is a much more elaborate treatment than usually occurs in natural resource management (usually we only talk about the last stage of evaluation—judging the value of the outcomes of some project—and ignore the other parts).

Fundamentally, however, the point of the chapter is that evaluation is an activity that helps make better decisions all the time.

## Items That May Come Up While Teaching This Chapter

In general, most students respond to this material with indifference. It appears to be academic, theoretical, and impractical. The chapter attempts to illustrate that these forms of evaluation are really just names given to things that we do regularly (i.e., formative = planning; process = accounting; summative = real evaluation), but it is sometimes a hard sell.

With this topic coming near the end of the book and the term in which it is taught, the students will be tired and perhaps a bit unwilling to undertake some of the more complex exercises. If a course had to be shortened somewhat to meet scheduling needs, this chapter could be covered quickly, without using the exercises.

## Supplementary Activities

Extra activities for evaluation are difficult. Because studying real evaluations requires studying the subject of the evaluation as well, it is difficult to suggest using some real evaluations as examples. Even acquiring copies of evaluations that have been conducted can be very difficult (e.g., most agencies and NGOs consider evaluations to be internal documents and do not spread them around). However, many groups conduct unsolicited evaluations of agency programs. Thus, The Nature Conservancy may have evaluations of some federal programs, the GAO regularly evaluates executive agencies, and the Congressional Budget Office conducts similar evaluations.

Students may also get involved if allowed to read or develop evaluation systems that affect them. Showing students a recent review of their department, conducted by the university, may stimulate interest. Also, giving students an opportunity to create novel systems for grading in a class may be useful. (The classification of evaluation presented in this chapter has an interesting message to students: Grades in classes are really process evaluations, not summative; they monitor progress, rather than outcomes.)

This is another area where inviting a representative of a state natural resource agency might be useful. Professionals in the planning division often have to deal with evaluation and can discuss how it works and why.

## *Conducting the Exercises*

### EXERCISE 12.1

#### How to Introduce or Conduct the Exercise

This exercise was designed to be done either individually or in small groups. Ask students to think about some evaluation activities they have been involved in, either as individuals or as part of a group. Students should be able to get into this quite easily, since they are evaluated more than about any group in society!

#### What to Expect When Students Do the Exercise

Students will generally have negative reactions to being evaluated. They will suggest that this a way to constrain them, control them, stereotype them. This is a perfect opportunity for asking students why these things exist—why do people think that things need to be evaluated? This should be a quick exercise, lasting only a few minutes.

#### Desired Outcomes for the Exercise

The exercise is designed to help students come to the conclusion that the purpose of evaluation is to help, not punish.

#### Points to Be Made Based on the Exercise

- Evaluation is hard, and it often hurts.
- We usually feel different when we are being evaluated ("they are stupid and unfair") and when we are evaluating (we are very willing to be harsh); how one feels depends on which side of the activity one is on.

### EXERCISE 12.2

#### How to Introduce or Conduct the Exercise

See the general instructions for the scenario exercises. This exercise is designed for students to think about the differences between objective and subjective data and when they would use either or both. It may be useful to introduce the exercise by giving students some familiar decisions that fall at the objective and subjective ends of the spectrum (e.g., picking a bank to establish a checking account and picking a spouse, respectively) and in the middle of the spectrum (e.g., choosing what car to buy). Buying a car is a very good example. Some students will think almost exclusively objectively (consumer ratings, gas mileage, cost), whereas others will think subjectively ("I want a red, fast convertible"). We all know, though, that we are influenced by both kinds of information.

If time for this exercise is short, doing only one of the two parts will provide most of the benefit. If a class is all using the same scenario, the second part (naming which topics would use mostly objective or subjective data) could be conducted as a full-class exercise.

### What to Expect When Students Do the Exercise

Students will be rather equivocal on this, in general. It requires them to think in ways that are much more explicit than usual. They may resist the exercise and need quite a bit of prodding to get going.

The ROLE Model example is the easiest of the three, because it is a simple task—naming a tract of land. The SnowPACT example may be the most familiar, because it asks students to make the kind of judgment that they are used to thinking about. The PDQ Revival is the most challenging.

### Desired Outcomes for the Exercise

The desired outcome is for students to recognize that both subjective and objective data are useful, but that both types are usually at play in any decision. Specifically, students should be able to understand the difference between objective and subjective data, recognize situations in which one or the other type of data is most useful, and accept the idea that decisions are seldom based just on objective data and seldom on as much data as they might wish.

### Points to Be Made Based on the Exercise

- Decisions use both objective and subjective information, so do not expect decision makers in natural resources settings to just use objective data.
- The kinds of data used and weights put on them depend on the decision.

## EXERCISE 12.3

### How to Introduce or Conduct the Exercise

See the general instructions for using scenarios. This exercise asks students to develop the decision-making model for a formative evaluation. It is an open-ended exercise, intended to have students think about what would characterize a desirable ecosystem project. Students should be encouraged to think about all they have learned in Chapters 5–10. Suggest that they develop a matrix of some kind, perhaps following Figure 11.4.

This exercise will require substantial class time for students to think, review, and prepare ideas. However, it can be conducted within one laboratory and discussion period because it is a conceptual exercise.

### What to Expect When Students Do the Exercise

Students may get bogged down trying to produce perfect criteria for some aspect that they understand well. Encourage students to go at this in a stepwise fashion, first getting more general ideas for all the items in Figure 11.4 before getting specific on anything. The idea is to develop a normative model for the specific ecosystem project, not a highly refined decision-making scheme.

### Desired Outcomes for the Exercise

The desire outcome is that students collect their knowledge from throughout the class and apply it in this circumstance. Students should develop an extensive list of criteria for ecosystem projects, covering all three aspects: ecological, socioeconomic, and institutional. Their list should include things that relate to the fundamental nature of ecosystems (e.g., cross-boundary, community-based, adaptive, stakeholder involved).

### Points to Be Made Based on the Exercise

- The characteristics they have written really define how ecosystem work differs from routine resource management.
- This is not just about ecology; it is also about people. It is a manifestation of the success triangle in Chapter 10.

## EXERCISE 12.4

### How to Introduce or Conduct the Exercise

This is a process-evaluation exercise. Students need to play the role of the folks who are responsible for making sure that their money is being spent well—that is, they are the responsible stewards for the resources their NGO is investing. This should be a relatively quick exercise that can be done in one sitting.

This exercise uses a general situation, rather than a situation based on the scenarios. We did this because some of the details of implementation are needed (funds, timing, specific project elements), and we did not want to bog the text down with those details for three examples. But the exercise may be more meaningful if the basis is a specific example that the instructor creates to match the kinds of details in the exercise as written.

### What to Expect When Students Do the Exercise

Some students will be bored by this exercise, partly because they may not have the experience to appreciate its importance and partly because they will consider it "bean counting." It *is* bean counting—and, therefore, is one of the first things they will run into when they get on the job. Nonetheless, it should be conducted as a quick, illustrative activity rather than something requiring great detail.

Other students, however, will really dive into the exercise, because it is an opportunity for them to set the rules, as rules have been set for them before. A group that produces many, strict, and detailed rules can be used as an example of how burdensome these exercises can become.

Students may also find this a place to complain about the punishment aspect of evaluation. This is a good time to discuss how the performance of one person or team influences other people and teams—and influences the ability to do good things for the ecosystem. In the example in the exercise, a team that did not get their work done or spent the money poorly would be causing the degradation of stream resources elsewhere.

### Desired Outcomes for the Exercise

The desired outcome is for students to be able to develop a workable process evaluation, at least in concept. The solution should develop criteria for each of the five general areas of process evaluation listed in the chapter. Students also need to consider how they would use the data for project adjustment decisions.

### Points to Be Made Based on the Exercise

- When you are in charge, you start asking for some pretty stringent reporting information and schedules.
- Somebody has to do the bean counting, and we all have to deal with it.

## EXERCISE 12.5

### How to Introduce or Conduct the Exercise

See the general description about using the scenario exercises. This is a major exercise in the chapter, covering summative evaluation. However, because it relates to a specific example, it can be completed in a relatively short time and in one class period.

### What to Expect When Students Do the Exercise

Students will enjoy this exercise because it lets them think very broadly about a particular situation. Because the nature of summative evaluation is expansive and research-like, this will appeal to students' desire to talk about consequences of action that go beyond the immediate.

### Desired Outcomes for the Exercise

The desired outcome is for students to create a comprehensive summative evaluation. Students should develop a list of 10–15 questions that cover all three aspects of concern (ecological, socioeconomic, institutional). Specifically, students should be able to distinguish between the kinds of questions appropriate for formative, process, and summative evaluations; understand that summative evaluations require both objective and subjective data; understand that summative evaluations require pre- and post-activity data; and relate summative evaluation to both scientific research and adaptive management.

### Points to Be Made Based on the Exercise

- Evaluating outcomes raises many interesting questions.
- Answering those questions is usually very difficult, time-consuming, and expensive.
- Evaluation often gets ignored because of these factors.

## *Teaching Devices to Help Illustrate the Chapter*

- Using easy examples to illustrate evaluation is a useful way to get the idea across. Box 12.1 can be used as a format for developing a complete evaluation set for a project that would be familiar to students. For example, a grant program to support activities of student groups in their college would be very familiar, and some students may actually have been on committees that make those decisions. They have probably thought about this some, and asking them to develop formative, process, and summative evaluations would be meaningful.
- Box 12.2 talks about the structure of state fish and wildlife agencies. Your state fish and wildlife agency could be evaluated according to that model and serve as the basis for a discussion about normative models.

## *Sample Test Questions*

1. Describe the uses and characteristics of each of the three styles of evaluation.
2. For a given situation (provide the situation), write a normative model that could be used for a formative evaluation.

## *Supplementary Web Sites*

- http://www.ag.ohio-state.edu/~ohioline/cd-fact/1261.html Information about conservation easements is available from The Ohio State University.
- http://www.rl.fws.gov/yreka/KRBFRP/ Information about the Klamath River Basin Fisheries Restoration Program is available from the U.S. Fish and Wildlife Service.
- http://gao.gov Publications from the Government Accounting Office deal with evaluations.
- http://www.cbo.gov The Congressional Budget Office likewise has information relevant to evaluations.

# Index